D0129677

THE HIDDEN IVIES

ALSO BY HOWARD GREENE AND MATTHEW GREENE

Greenes' Guides to Educational Planning: The Public Ivies

Greenes' Guides to Educational Planning:Inside the Top Colleges

Greenes' Guides to Educational Planning: Making It into a Top College

Greenes' Guides to Educational Planning: Presenting Yourself Successfully to Colleges

Greenes' Guides to Educational Planning: Making It into a Top Graduate School

THE HIDDEN IVIES

THIRTY COLLEGES OF EXCELLENCE

Howard R. Greene, M.A., M.Ed.,
and Matthew W. Greene, Ph.D.

HarperResource

An Imprint of HarperCollinsPublishers

THE HIDDEN IVIES. Copyright © 2000 by Howard R. Greene and Matthew W. Greene. All rights reserved. Printed in the United States of America. No part of this book may be used or reproduced in any manner whatsoever without written permission except in the case of brief quotations embodied in critical articles and reviews. For information address HarperCollins Publishers Inc., 10 East 53rd Street, New York, NY 10022.

HarperCollins books may be purchased for educational, business, or sales promotional use. For information please write: Special markets Department, HarperCollins Publishers Inc., 10 East 53rd Street, New York, NY 10022.

FIRST EDITION

Designed by Jackie McKee

Library of Congress Cataloging-in-Publication Data
Greene, Howard 1937-
    The hidden ivies : thirty colleges of excellence / Howard R. Greene and Matthew W. Greene.
       p. cm.
    Includes index.
    ISBN 0-06-095362-4
    1. College choice—United States—Handbooks, manuals, etc. 2. Universities and colleges—United States—Entrance requirements—Handbooks, manuals, etc. 3. Universities and colleges—United States—Admission—Handbooks, manuals, etc. I. Greene, Matthew W., 1968- II. Title.

LB2350.5 G74 2000
378.1′61–dc21   00-043157

08 09 10 RRD 10 9

*A university should be a place of light,
of liberty, and of learning.*

—Benjamin Disraeli

*Training is everything.
The peach was once a bitter almond;
cauliflower is nothing but cabbage
with an education.*

—Mark Twain, *Pudd'nhead Wilson*

CONTENTS

ACKNOWLEDGMENTS]

One of the many blessings of having been grounded in a traditional liberal arts education is the exposure we received to many of the great social and literary thinkers like Samuel Johnson, the eighteenth-century writer.

Johnson said that, "Knowledge is of two kinds. We know a subject ourselves, or we know where we can find information upon it." As counselors and writers in the educational field, we possess a fair amount of knowledge about higher education, individual colleges, and the planning process for the transition from high school to college. But there is so much that we cannot know by ourselves. Fortunately, we know where to turn for further information and insight. This book could not have been produced without the generosity of spirit and cooperation of the many individuals who have much to teach us: the administrative leaders of the colleges represented in this book and the students who are the beneficiaries of the superb education provided by these institutions. In all cases, their enthusiasm for their respective colleges and the honest portrayal of their strengths and special qualities make this book meaningful. To them all we tip our faded mortarboards and thank them for giving so freely of their time and thought.

To our editor and publisher, Diane Reverand, we can only declare our gratitude for her vision of the importance of higher education and the college admissions process, and her confidence in us to cast a brighter light on a complex and sometimes bewildering subject. To Matthew Guma, who has served a critical role as editor, questioner, implementer, and enthusiast, we say thank you for your engagement and support.

We are delighted, as well, to thank our associate at Howard Greene and Associates, Doris Forest, who knows instinctively when it is necessary to convert concepts and ideas that have a way of floating in space into an action plan. Her ability to help us move forward in an orderly fashion and to connect with the sources of information in a timely manner has been essential to the completion of this task.

THE HIDDEN IVIES

INTRODUCTION]

Our mission in writing this book for students and parents is to create greater awareness of the small, distinctive cluster of colleges and universities of excellence that are available to gifted college-bound students. We would be surprised, indeed, to encounter a student or parent throughout the country or outside the United States who has not heard of the exclusive band of elite institutions known as the Ivy League. They have become, in our modern economic universe of marketing, the ultimate brand name for quality education and prestige. It is a challenge not to become intoxicated with the thought of graduating from one of these famous schools. Each year these eight institutions—Harvard, Yale, Princeton, Dartmouth, Brown, Columbia, Cornell, and the University of Pennsylvania—are flooded with applications from thousands of fully qualified individuals who, despite the fact that they have done everything one could ask of them as students, will not be invited to enroll in any of these schools. Other universities such as Stanford, MIT, Cal Tech, Virginia, Duke, Georgetown, and the group historically known as "the Little Ivies" (including Amherst, Bowdoin, Middlebury, Swarthmore, Wesleyan, and Williams) have scaled the heights of prestige and selectivity and also turn away thousands of our best and brightest young men and women. This is the present situation, and it is one that will be with us for a number of years down the road.

Parents can prepare for their role of supporter and guide to their children as early as junior high school, since numerous surveys have revealed that they are the number-one influence in their children's choice of college. The more you are armed with information on the breadth of top college choices to aim for and what is required for admission, the better able you will be to serve your son or daughter. Students can profit by reading—in the early years of high school—about the factors that make for an outstanding college and which ones might match up best for them. If they then follow the essential steps we outline, they will better their odds of being accepted to the college of their choice.

Let's go directly to the heart of the matter: There are two dynamics at work that have combined to intensify the competition for admission to the best colleges. The number of high school graduates of exceptional ability and motivation who have set their sights on gaining the very best college education they can have is rising dramatically each year, as is the cost of receiving such an opportunity. Admissions deans at the selective colleges are shaking their heads in astonishment at the academic and other achievements of applicants to their schools. Never before have they seen so many young men and women who have attained such outstanding records in their high schools, performed at such high levels on the SAT and ACT, and garnered so many awards for their artistic, athletic, and intellectual capabilities. They worry, just as parents do, about the constant rise in tuition, fees, and room-and-board expenses. It has become an ongoing contest of increasing financial aid in order to stay ahead of the annual increases in the cost of a selective college education, which, at the top end, is now in the $33,000-per-year range.

In 1999, 1.3 million students took the SAT and another 1.2 sat for the ACT. Over 250,000 applications were received by the thirty Hidden Ivies and the eight Ivy League institutions alone in 1998. A great many of the best students are seeking guidance on identifying the best colleges and how to get into them. We have written this book to guide you on both counts. Parents and students are in a continual frenzy over the preparation for and admission to a top college today. The cost factor and the perception that the prestige factor will determine the quality of life their children will achieve has turned them into their own research-and-guidance teams who seek information from every source. *The Hidden Ivies* fills a large gap in the college admissions literature by providing the thoughtful researcher with descriptive, responsible, and insightful discussions of an exciting group of colleges and universities toward which many families are increasingly turning.

Both students and parents are greatly concerned about fulfilling the dream of attending a top college or university. As the number of outstanding candidates to the most selective schools continues to escalate, the competition for admission becomes more intense since the number of available places has not increased over time. To make matters worse, in the past several years virtually all of the Ivies and their counterparts have experienced a higher yield—the percentage of admitted students who choose to enroll—of accepted candidates than anticipated, which leass to overenrollment and which drives down the number of applicants who can be accepted in subsequent years! This is, in fact, one of the virtues of the top colleges: their determination to retain their historic enrollment in order to deliver outstanding teaching and a sense of community that has much to do with the learning experience.

Every day in the offices of Howard Greene and Associates, we meet with families who are embarking on the search for the best college for their children. One of the very first questions we are asked by many parents concerns the odds of their son or daughter getting into one of the Ivies. The next question, invariably, concerns what other top colleges to consider if this is either an unrealistic or uncertain goal. As the cost of a college degree continues its relentless climb, the name recognition of an institution becomes extremely important to the educational consumer, both to certify that the benefits of such a program are worth the high sticker price and to assure acceptance to graduate school and a good job at a later date. A commonly heard refrain of concerned parents is that they have so little real knowledge of, let alone insight into, the top colleges beyond the prestigious ones. Here is the point at which thousands of ambitious students need direction to these other great institutions, as well as greater understanding of their particular styles of teaching, the areas of academic and faculty strengths, and the predominant social and learning environments.

We are not surprised by the public's lack of awareness of the many other colleges and universities that stand for excellence and may, in fact, be the most appropriate educational choice for many students. With such a vast array of public and private colleges available, the consumer can easily be overwhelmed in sorting the wheat from the chaff. Highly sophisticated marketing and image building by the majority of colleges make for quite a challenge in differentiating the great from the mediocre. What you will find here is a window into those colleges and universities that we consider to be the Hidden Ivies, those institutions that meet all the criteria we believe set the standard for excellence. Any ambitious and talented high school student would do well to enjoy four years at any one of these colleges. We encourage you to launch your college search with an open mind, a spirit of adventure, and the confidence that you will ultimately enjoy your collegiate years and be prepared to move on to a fulfilling personal and professional life. One of our students, when asked in our survey what we should have asked, said: "If this book is called *The Hidden Ivies*, you should have asked why students at these schools chose not to attend the real Ivies." We think we have answered this question through our discussion of what makes an Ivy an Ivy, the criteria for selecting a school of excellence, the value of the liberal arts education at the Hidden Ivies, and the individual college descriptions.

What makes *The Hidden Ivies* unique in the library of college guidebooks is that we bring to the reader the voices and experiences of the educational leaders as well as those of students at the institutions under review. We are gratified by the spirit of generosity on the part of both groups. They have shared with us their vision of their institution and its essential qualities through lengthy interviews and written responses. Too many guidebooks attempt to capture the definitive picture of a college through a few interviews with students or a brief visit to the campus. We believe strongly that the true nature of any educational enterprise is determined by its leaders, who are responsible for setting its goals and policies and then implementing them. Present and past students are in a position to articulate whether these goals are met by what they gained from their time on campus. We have followed many of the students we counseled into the colleges reviewed here so that we have a sense of their personal growth and satisfaction over time. This confirms our sense of what type of student and personality will thrive on a particular campus.

We believe that our professional experience in counseling college-bound students and researching colleges for more than thirty years enables us to determine the essential components that create an outstanding educational experience. By identifying a cohort of thirty colleges of exceptional merit and an additional group of wonderful institutions that merit recognition, we want to help ambitious students select the best available opportunity that fits their interests, personality, learning style, and goals for their future.

The degree of enthusiasm for their respective colleges expressed both by students and the institutional leaders serves as confirmation of our selection of the colleges included in *The Hidden Ivies*. We are deeply indebted to the individuals who took precious time from their heavy schedules to cooperate in this review. The opportunities for intellectual and personal development they provide for present and future generations of gifted young men and women stands as the ultimate reward for their labors.

In our book *Inside the Top Colleges*, we describe the fundamental reasons that underlie the intense drive to enroll in the eight Ivy League and twelve of their peer colleges today and the impact on the student who actually gets to experience life on one of these campuses. To make the point, of the 1.4 million students who entered college in the past five years, less than 5 percent were enrolled in the fifty most selective public and private colleges. Of the 1.3 million high school seniors who took the SAT this past year, only some 30,000 scored over 700 on the verbal section of the test. One-third of these high scorers are enrolled in only ten colleges in America. Almost half are enrolled in one of the thirty most selective colleges. Ten prominent institutions—the eight Ivy League schools and Stanford and MIT—enrolled a mere 1 percent of all entering freshmen, yet they represented one-third of the highest academic achievers and test-takers in the nation.

Despite the fact that not every talented student can gain admission to so small a segment of our vast higher educational system, the dream continues in the minds and hearts of a great many parents from the moment their children are born. Eventually it becomes the overriding ambition of tens of thousands of students as well. Success for both the parent and the child is measured often by the prestige of the college to which one is admitted. The perceived reality in our intensely competitive society is that a first-rate education from a prestigious institution confirms the quality of the diploma and thus drives more and more people to set their sights on making it into one of the most elite institutions.

In interviewing more than four thousand enrolled students in the top colleges and universities for *Inside the Top Colleges*, we received confirmation that one of their primary reasons for choosing to enroll was the prestige and name recognition attached to their school. They assumed that the quality of education they would receive and their prospects for their future positively correlated with the prestige factor. While we brought to light some of the negative factors that are to be found in these great universities, we did find a high overall level of student satisfaction and appreciation for the education they received. The major benefits were the expertise of the faculty, the exceptional resources available for learning in and out of the classroom, and, of equal or greater importance, the talents and energies of their fellow students. We will review some of the less desirable factors that influenced their experience, such as access to faculty and large classes, since the cluster of select colleges in this book meets these issues successfully.

Let us discuss up front a question in the minds of most families: Is a college degree worth the huge financial investment and commitment of time that is required? The present costs of a four-year education averages $12,000 at the public universities and $20,000 at the private colleges each year. The most elite private colleges, however, now have an average cost of $30,000 per year, including room and board and student fees. This represents an increase of 234 percent since the early eighties, while median household income rose only 83 percent! But keep in mind that these schools have exceptional levels of scholarship support because of their commitment to enrolling gifted students of all backgrounds. They also have the endowments to meet this goal. The phrase repeated constantly by the admissions staff, put simply, is: Ignore the sticker price and concentrate on your ability to qualify for admission, since we have such generous financial assistance for those who demonstrate need.

To return to the question of whether a family can justify the high costs of a college degree, the answer is a resounding yes! In the early part of the nineties college-educated men earned an average salary of $45,000, while men with only a high school diploma earned $27,000. Yes, the inequity between men and women continued through this same period. However, college-educated women earned on average $32,000, while female high school graduates averaged $18,000.

College graduates not only earn more over the course of their lifetimes but are significantly less at risk to be unemployed. In the decade just concluded, only 4 percent of college graduates were unemployed involuntarily, compared to more than 8 percent of high school graduates. These are quantifiable numbers, of course, and do not account for the personal and intellectual enhancements that can affect the quality of one's life. We argue here that the benefits of attending a Hidden Ivy or a college of similar character and quality are significant in terms of teacher and peer standards, campus resources, a definite sense of campus community, and intellectual challenge. All of the factors contribute to a student's lifelong learning experience. We also make the case for the continuing value of a liberal arts education, one of the defining strengths exhibited by the Hidden Ivies.

As we enter a new century, we will be confronted with a rising tide of students who will swell college campuses. Elementary and high school enrollment will surpass the peak level reached in 1977, the summit of the post World War II baby boom, increasing to some 53 million students. Enrollment in grades 9 through 12 will increase by 21 percent. High school graduates will grow by 17 percent, to 3 million, and the number of college students will grow to 15.5 million. Space availability in the most selective colleges and universities will remain static at the same time that more and more aspiring scholars and future professional leaders in all fields will vie for these limited places. In a society in which access to the best career opportunities is often linked to the name factor of the college and graduate diploma, the race for admission to the very best of our colleges will continue at a pace unimagined a generation ago. Admissions deans at the top-end colleges and universities have expressed both their surprise and delight at the acceleration in the numbers of applications from outstanding high school seniors, but have noted the difficulties associated with their not having enough places to offer all those talented applicants.

There is, however, another side to the condition of America's universities. A recent report from the Carnegie Commission on Education describes in negative terms the conditions of teaching and learning in the top research universities, which underscores the importance for students to have accurate and relevant information in making the right choice of college. The Carnegie study confirmed what we as professional counselors and educators have known right along since we speak with undergraduates across the country: The great majority of universities are growing their enrollments at the same time that they are in desperate need of containing or reducing the costs of operating their institutions.

Economic efficiency is all too easily translated into larger classes to maximize the time and talent of the tenured faculty, the use of graduate students who are pursuing their own studies under great pressure at these research universities, and, what is fast becoming a permanent fixture on campuses, part-time instructors who spend infrequent time with their students because they typically have to hold such positions at several colleges to eke out a living. Administrators are spending much of their time seeking cost-efficient methods of teaching with modern digital technology. Going down this path involves even greater separation of teachers from their students, who find that they have less and less interaction in the classroom with the experienced and well-known faculty and certainly little or no opportunity to seek their guidance outside the classroom. Many of the best of the public universities have recognized these issues and have instituted such programs as honors colleges, residential learning programs, and special honors or critical-thinking curricula in order to emulate their smaller, liberal arts brethren. They have found these special programs to be attractive to large numbers of outstanding high school students who find the combination of an elite residential learning program at state-school prices too appealing to turn down. This is a trend in the right direction. However, today only a relatively small number of colleges have the financial resources, a sense of their historic mission, and the determination to limit their growth so that they can provide their students with the essential ingredients of an excellent education. Further on we discuss the best of the public universities and the advantages they have to offer an ambitious student. Here, however, we put our focus on the small and medium private colleges and universities. These are the Hidden Ivies.

Inside the Top Colleges was a groundbreaking study of twenty of the most elite colleges in America in that it surveyed in great detail the experience of the students who attend them.* The knowledge of the many strengths and some of the weaknesses of student life in these major colleges will serve as a background in describing the Hidden Ivies. As we found, every institution has its strengths and its weaknesses. We thought it important to college-bound students to uncover those factors can which result in a highly positive or a potentially negative college experience.

---

*The twenty colleges and universities included in *Inside the Top Colleges* (formerly titled *The Select*) are: Brown, Columbia, Cornell, Dartmouth, Duke, Georgetown, Harvard, Johns Hopkins, MIT, Northwestern, Princeton, Stanford, University of California, Berkeley, University of Chicago, University of North Carolina, Chapel Hill, University of Pennsylvania, University of Wisconsin, Madison, Wesleyan, Williams, and Yale.

We begin this book with a discussion of those primary factors that give an Ivy League college its level of excellence and thus its reputation, as well as those elements that could be improved upon. We then identify thirty institutions that we believe merit similar consideration. We have included our top choices of those independent residential liberal arts colleges and small universities that we believe match up with the Ivies. Note the emphasis given to the residential nature of all thirty institutions because of the profound and lasting impact of four years spent interacting continuously with a broad variety of students and faculty who are present to teach and advise. The further students get into their college years, the greater the value they place on the interaction with their peers and teachers. This, rather than location, enrollment, nature of the food and dormitories, and the availability of alcohol, as examples, will determine their affective and intellectual development and level of satisfaction.

The colleges and universities we have selected for inclusion all provide a shared community experience, a virtual living/learning laboratory for intelligent young men and women at a most impressionable stage of their lives. They are likely to acquire the following critical skills or instincts: cooperation, leadership, sharing, deferring appropriately to the needs and sensitivities of others, social and intellectual discourse, conflict resolution, collaboration, mentoring, self-reliance, coping with competition, independence, control of one's emotions, delayed gratification, organizing and planning, finding one's place in the immediate and the larger community, appreciating religious and cultural differences, making decisions that count, not procrastinating, not blaming others for one's actions, and learning the truth that intelligence without character, personal integrity, and a working set of values can be a dangerous thing.

We hasten to add that this is not an exhaustive list. There are many outstanding colleges and universities that could have been included. However, we have cause to place these particular institutions in the top group. We recognize another cluster of outstanding colleges and universities for their benefit to students in search of a great educational experience in the Appendices. We know from our many years of experience as counselors how valuable this guidance can be for concerned students and their parents. Just as "Ivy-bound" students must broaden their college search to include some of the Hidden Ivies and other schools of excellence, those applying with confidence to the Hidden Ivies must expand their view of potential colleges that will meet their aspirations. We hope that our discussion of the Hidden Ivies and the criteria that make them top colleges will serve as a template for students and parents for evaluating the many other options of higher learning available to them.

We hope also that this book will prove of particular value to the increasing number of college-bound students of color and from nontraditional backgrounds. Students of African-American, Hispanic, and Asian heritage will soon represent 20 percent of our school population. These young men and women, like all ambitious students, deserve the opportunity to know which educational direction they can take to ensure a future of security and meaningful engagement in the affairs of our country. As affirmative action policies are challenged, and even dismantled in many of our great public universities, many of the private colleges we discuss in *The Hidden Ivies* continue to be proactive in offering disadvantaged students access to excellent educational opportunities. Their financial strength is a vital source of support for the talented, disadvantaged candidate. We heartily endorse their commitment and would like them to be identified through this guide.

We attempt in Chapter 1 to educate college-bound students on the value of acquiring a liberal arts education. While it appears a contradiction in terms at first glance, exposure to the full range of intellectual disciplines—from the social sciences, humanities, and the arts to sciences and mathematics—is vital for a successful professional future when the prevailing notion is that a college education is solely a training ground to prepare graduates for employment. From the constant flow of questions from parents and students alike, we recognize the importance of explaining the value—personal, intellectual, and professional—of spending four critical undergraduate years learning how to think critically, read and analyze advanced ideas and concepts, think for oneself by developing the art of disciplined evaluation, and discover the pure joys of art, music, literature, science, and the many other fields one will encounter in a first-rate liberal arts college. The emphasis on grounding their education with an ethical and moral awareness is implicit in the teaching of a majority of the faculty. It is not a mere coincidence that all of the private colleges selected as Hidden Ivies offer solely or primarily a liberal arts education at the undergraduate level. This is true of all of the Ivy League institutions, to make our point.

In Chapter 2 we detail the criteria that we believe make a college or university an outstanding institution. Careful attention is given to each college's mission statement because this is the vision or blueprint, if you will, by which the leaders decide on the programs, policies, and content of the curriculum and nature of teaching. You will learn how so many institutions of excellence can define their environments and learning approaches in differing ways. We describe common elements, such as the ratio of the size of the endowment to that of the student body and what this means for educational excellence, the educational attainments of the faculty, the physical facilities for learning (library, science, visual and performing arts, and athletic programs and facilities) in relation to the student population, the resources for advising and learning support, graduate school admission in all of the major disciplines, and the response by the administration to student needs. We share with the reader input from present undergraduates on those elements that most determine the quality of experience in the individual college descriptions.

We then provide the reader in Chapter 3 with a summary of the overall character of each of our cluster of Hidden Ivies as gleaned both from the input of the administrators and the students surveyed. We also take advantage of our many years of experience with students attending these colleges and an awareness of what kind of individual would function best in each case. Information on cost, requirements for admission, and special qualities or talents that are sought by the admissions committee will be included. This is not intended to be still another standard guidebook or rankings list, filled with mere facts and information provided mainly by the institutions themselves. The descriptions address the unique personality of each college, primarily through the voices of the students, who we believe speak for a majority of the student body, and the key administrators who generously shared their views and expertise with us. The colleges will be discussed in terms of the criteria we have determined are of paramount importance in judging an institution of higher education.

In gathering the data for *Inside the Top Colleges*, a detailed questionnaire was completed by students on all twenty of the elite campuses. This time, however, we surveyed the college leaders of each school as well, from the president to the deans of faculty and admissions, the provost, and the professional support staff. A sample questionnaire that students and administrators completed is included in the Appendices. Many college counselors who have years of experience in guiding students to the top colleges also shared their recommendations for the Hidden Ivies list with us.

Few studies of colleges and universities have directly asked both the student, who is the primary consumer, and the educational leaders responsible for providing the education specific questions regarding the nature and quality of their experience other than superficial and anecdotal queries. We believe that we have garnered important information and insights that will prove of great value to college-bound students, their parents, counselors who are engaged in guiding their young charges, and the college leaders and teachers who run these institutions. There is no other book in the marketplace that focuses exclusively on those colleges, public and private, that are equal to or almost as impressive in all respects as the Ivy institutions.

In Chapter 4 we conclude *The Hidden Ivies* with our list of key recommendations for successful admissions as well as a "Parent to Parent" and "Student to Student" section that lets those who have recently gone through the college search process share their suggestions with the inexperienced parent and college-bound student. We detail tips on what they did to assist their children, what mistakes they would now avoid, what they have found makes for a right or wrong choice of school, and what misperceptions they had that they would alert you to. We encourage you to keep in mind that all of the colleges of excellence included here will lead you to where it is you think you want to go. It is the choice of routes and the nature and style of travel you prefer that will differ. This is what we will help you to discover. So, let us begin our journey through the land of the Hidden Ivies.

ONE The Liberal Arts:
WHAT IS A LIBERAL ARTS EDUCATION AND WHY IS IT IMPORTANT TODAY?

Every session we have with parents and students on college planning inevitably leads to this question: "Why should my child go to a liberal arts college? How will it prepare him/her for a job or career? Why not jump right into a specialized major?" In this chapter, we answer that question. We have found that there is a great deal of misunderstanding and a general lack of information about the nature of a liberal arts education, its value in society, and its role in preparing students for graduate programs and careers. Many students are unaware of the differences between a "college" and a "university," between a graduate and an undergraduate education and degree, and between degree programs specializing in technical, business, arts, or other fields and those offering or demanding study across the arts and sciences disciplines.

We defend the value of a liberal arts education, building on the work of many prominent scholars who have argued that an education spanning multiple academic disciplines and requiring that students learn core concepts, methods, and content builds unparalleled strengths in reasoning, understanding, and communication, preparing students for any academic or professional challenge they may choose. At the same time, we avoid for the most part the disagreements among "liberal" and "conservative" thinkers on necessary reforms in liberal arts education and on whether changes over the past several decades have been "good" or "bad." However, we must add that in our minds, liberal arts colleges have changed for the better and significant choice exists among these institutions to give students a great deal of leeway in determining which curriculum and environment best suits their needs and interests.

GOALS OF A LIBERAL ARTS EDUCATION

As Nathan Glazer recently stated: "Liberal education has meant many things, but at its core is the idea of the kind of education that a free citizen of a society needs to participate in it effectively."* In a complex, shifting world, it is essential to develop a high degree of intellectual literacy and critical-thinking skills, a sense of moral and ethical responsibility to one's community, the ability to reason clearly, to think rationally, to analyze information intelligently, to respond to people in a compassionate and fair way, to continue learning new information and concepts over a lifetime, to appreciate and gain pleasure from the beauty of the arts and literature and to use these as an inspiration and a solace when needed, to revert to our historical past for lessons that will help shape the future intelligently and avoid unnecessary mistakes, to create a sense of self-esteem that comes from personal accomplishments and challenges met with success.

- Think and problem-solve in a creative, risk-taking manner.

- Express ideas and feelings in organized, logical, coherent, descriptive, rich language both orally and in writing.

- Analyze, organize, and use data for meaningful solutions.

- Develop the capability of setting goals with appropriate information and research and then achieve these goals with proper means.

- Help define a personal-value and ethical system that serves throughout life in making the challenging decisions one will face.

- Have the capacity and instinct to work in a cooperative, collaborative manner with others in one's professional and community life.

These are ambitious goals! How different colleges and universities achieve them reveals variations in educational philosophy, institutional personality and history, and particular social and academic strengths and missions. All the liberal arts colleges share a commitment to disciplinary and student diversity, intellectual and otherwise. To varying degrees, these colleges require students to pursue courses in key academic subject areas, some with more specificity, others with a great deal of freedom, in order to expose students to multiple areas of knowledge, diverging perspectives on the world, and different paths to scientific, ethical, social, and humanistic understanding.

Content is important, but so are process and style. Liberal arts colleges may expect students

*Glazer, Nathan. "After the Culture Wars," *New York Times Review of Books*, 7/26/98:6.

to master a core body of knowledge, including Western and, increasingly, non-Western master-works in fields ranging from physics to music to government, comparative literature, history, and language. Students will build on their secondary school education by majoring in one or more specific areas of knowledge (academic "disciplines" or "fields"), but will pursue areas of interest within key general academic areas: the physical sciences; mathematics; the humanities (history, English and other languages, visual and performing arts, and so forth); and the social sciences (political science, sociology, etc.). So-called "cross-disciplinary" courses of study are offered in such areas as women's studies, African-American studies, environmental studies, and social psychology. Students will be exposed to a wide range of subjects that they may not have encountered previously: anthropology, genetics, philosophy, criminology, economics, engineering sciences, religion, education. But all of this will be in the context of a broad-based approach to learning. One cannot graduate from a liberal arts college without having experienced coursework in a multitude of subject areas. The goal: an intelligent and "well-educated" student who can converse knowledgeably about a wide range of topics and who has learned how to learn about anything under the sun.

Thus, style and process undergird a liberal arts education. Students learn how to think, approach problems, write, present information intelligibly, and make coherent arguments in their field of choice and others they may encounter. A liberal arts education challenges students' conceptions and pushes them to ask difficult questions, question established answers, and develop their own arguments through logical reasoning and the discovery of new understandings. A liberal arts education strives to help a student specialize in at least one particular area, but also to see and make connections among multiple fields of inquiry. As Ernest Boyer, a former president of the Carnegie Foundation for the Advancement of Teaching, has argued, traditional research designed to promote the advancement of knowledge should be complemented by the "scholarship of integration," which makes connections across disciplines; the "scholarship of application," which concentrates on the interrelationship between theory and practice; and the "scholarship of teaching," which both educates students and attracts them to the academic world. Such a view of scholarship clearly relates to the goals of a liberal arts education.

Learning How to Learn: The Luxury of Time

Alan Ryan writes: "At its best, liberal education opens a conversation between ourselves and the immortal dead, gives us voices at our shoulders asking us to think again and try harder."*How many of us, academics or not, would not relish the notion of taking four years of our lives to keep open that conversation, enjoying what we call the luxury of time, to think, to make connections, to question, and to learn? Howard Bowen, a highly respected

*Ryan, Alan. *Liberal Anxieties and Liberal Education* (New York: Hill & Wang, 1998). Quoted in Glazer, Nathan. "After the Culture Wars," *New York Times Review of Books*, 7/26/98:6.

teacher/researcher of higher education, writes in a reprint of an earlier important study:

> As compared with others, college-educated people on the average are more open-minded toward new ideas, more curious, more adventurous in confronting new questions and problems, and more open to experience. They are likely to be more rational in their approach to issues. They are more aware of diversity of opinions and outlooks, of the legitimacy of disagreement . . . They are less authoritarian, less prejudiced, and less dogmatic. At the same time, they are more independent and autonomous in their views, more self-confident and more ready to disagree. They are more cosmopolitan.*

Bowen's point is that attempts to measure the value-added benefit of a college education often consider the wrong issue. The major benefit of a liberal arts education is that it will produce the kinds of educated leaders that will benefit our economic, political, social, and family lives. Such leaders include Carleton S. Fiorina, a woman recently named chief executive of Hewlett-Packard. Breaking gender barriers in leadership in the important telecommunications and computer industries, Ms. Fiorina has an undergraduate degree from Stanford University in liberal arts. Her fields? Medieval history and philosophy.+

John Wooden, the great basketball coach at UCLA, once quipped, "It's what you know after you learn everything that counts." A liberal arts education, particularly one obtained in a residential college setting, seeks to provide that learning experience and that sense of knowing a lot, but also knowing what you do not know. "We go to college," the poet and teacher Robert Frost said, "to be given one more chance to learn to read in case we haven't learned in high school. Once we have learned to read, the rest can be trusted to add itself to us." As with the majority of his observations and commentary, Frost is full of irony in his view of the purpose of the liberal arts experience. His message, as true today as it was many years ago (when Frost was at Dartmouth College and Amherst), is that reading intelligently and analytically, with a critical mental eye, will enable one to carry on his/her education for the remainder of a lifetime. In a sense, once one has "learned to read" in its broadest meaning, he/she is prepared to go out into the world. It is the definition of reading that counts.

Ernest Martin Hopkins, a former president of Dartmouth College, characterized in his 1929 convocation address to the entering class the view of the essential qualities of the liberal arts college: "The liberal arts college is interested in the wholeness of life and in all human activity . . . It is characterized as liberal because it recognizes no master to its limit to seek knowledge and no boundaries beyond which it has not the right to search. Its primary concern is not with what

*Bowan, Howard R. *Investment in Learning: The Individual and Social Value of American Higher Education* (Johns Hopkins University Press, Reprint Edition, January, 1977).

+Markoff, John. "Hewlett-Packard Picks Rising Star at Lucent as Its Chief Executive," *New York Times*, 7/20/99:23–24.

men and women shall do but with what they shall be." The liberal arts education is the means by which outstanding young men and women will develop those skills and qualities of mind and spirit that will enable them to lead productive and valuable lives. This means not only for their own well-being, but also for the good of their families, communities, and the larger society. The colleges we have selected are among the leaders in higher education in preparing young adults to take their places as responsible and enlightened leaders in the world. To think critically and with a conscience, to be resilient in an accelerating world of technological, intellectual, cultural, and social change are critical skills for the individual and society's maintenance.

AN EXAMPLE OF THE LIBERAL ARTS IN ACTION

Pomona College's faculty has a ten-point list of what skills it teaches to help students succeed in later life through exposure to the arts and sciences curriculum. The college educates students to:

1. Read literature critically.

2. Use and understand the scientific method.

3. Use and understand formal reasoning.

4. Understand and analyze data.

5. Analyze creative works critically.

6. Perform or produce creative art.

7. Explore and understand human behavior.

8. Explore and understand historical cultures.

9. Compare and contrast contemporary cultures.

10. Think critically about values and rationality.

TRENDS IN THE LIBERAL ARTS OVER TIME

The liberal arts colleges in America are dynamic institutions, constantly, if slowly, evolving in reaction to their environment, the demands of students, parents, graduates, policy makers, and others, and the leadership of administrators and faculty. Important changes have taken place in the liberal arts and in post–secondary education over the past several decades, including diversification of staff, student body, and curriculum; provision of financial aid; accentuation of the continuing struggle between teaching and research; expansion of educational access; and pursuit of graduate degrees. We discuss these trends and others on the following pages.

Inclusion

Perhaps one of the most powerful changes in liberal arts colleges since the 1950s has been the expansion of educational opportunities for non-white, non-Protestant, non-wealthy, and non-male students. The policy of exclusion at the elite institutions based on racial, ethnic, religious, and socioeconomic factors is clearly a thing of the past. These colleges have redefined their missions and have worked assiduously to attract, retain, and assist a student body that better reflects American society.

In addition to doing away with formal, public, or hidden policies of exclusion, selective colleges have actively tried both to "change their image" and broaden their appeal and impact. They recruit students from "non-traditional" backgrounds by visiting their schools and writing to them. They hire special admissions officers of color, put together targeted informational materials, work with the College Board to identify talented test-takers, and form close associations with organizations like A Better Chance and Prep for Prep, which help students of color succeed and go on to college. Colleges have boosted financial aid resources and expanded efforts of alumni to identify appropriate candidates. They have hired multicultural advisers and staffed centers on campus to encourage tolerance and diversity and to support each student's needs. Of course, this is a continuing process, and colleges, not to mention American society, are learning about what it takes to promote and maintain diversity in a way that is successful for everyone. But a major shift in thinking has occurred so that not only do colleges prohibit and discourage exclusion, they see inclusion and diversity as essential elements of their educational missions.

Liberal arts colleges have also continued to diversify their faculty, who serve as role models and offer differing voices to expose students to multiple perspectives. In the past, class snobbery extended to the college faculty as well as to the students. Very few teachers were not white, male, Protestant, and educated from the same small band of institutions. Today, liberal arts colleges certainly cross-fertilize each other's faculties, but representation of multiple groups and viewpoints on the faculty has dramatically increased.

Not only are the faculty more diverse at the liberal arts colleges, but the courses they teach have broadened and fragmented. At most liberal arts colleges today, students can access courses in new areas of scholarship, including African-American studies, women's studies (gender studies), environmental studies, non-Western literature, ethics and science, and so forth. The "core curriculum" of the great works of Western literature, history, philosophy, mathematics, sciences, and the arts has splintered and branched. Few colleges have completely dismantled the "core." Most have found a middle path, combining "traditional" scholarship with exposure to new and alternative ("critical," "postmodern," "non-Western") fields of study.

Finally, we should mention the prominent role of athletics as a means of recruitment among disadvantaged and nontraditional groups over time. This was one of the first vehicles by which colleges identified talented students in nontraditional environments. While the Ivies still do not offer athletic scholarships, their need-based financial aid has allowed them to attract and enroll "scholar-athletes" regardless of ability to pay. The fact that at many selective liberal arts colleges some 50 percent of students are receiving some sort of financial aid indicates their willingness to promote socioeconomic diversity. Many potential students and families may still react negatively to the cost of private liberal arts colleges and the image they may still carry of being snobbish, elitist places. However, while these colleges do not have the natural diversity that many state universities likely attract, they have pursued diversity through active recruitment of various student populations. And the public universities have launched similarly active programs to attract students of color and other underrepresented groups.

Liberal Arts Colleges: A Small But Important Sector

Only 4 percent of today's collegiate student body is enrolled in residential liberal arts colleges, yet these institutions remain a vital force in setting high standards of teaching and intellectual rigor. Many larger institutions have modeled their undergraduate liberal arts programs on the traditional residential college model. What also appears to be true is that the small band of such colleges carries a disproportionate influence in higher education. The recent Carnegie Report on teaching, as well as many prominent educators, urge the large universities to follow the residential colleges' emphasis on intellectual standards and teaching and interaction with students. The growth in honors programs and residential colleges within larger universities testifies to this fact.

What drives the degree of excellence of education at the liberal arts colleges is the caliber and commitment of the faculty. Endowment building and alumni support are the keys to creating and sustaining a first-rate faculty. This is the essence of the commitment of these colleges. One does not find an inexperienced graduate teaching assistant or an adjunct instructor, or a large lecture hall outfitted with a large video screen. An education in the liberal arts fits hand-in-glove with a teaching style that emphasizes faculty-student interaction, discussion-based learning, and opportunities to practice writing, speaking, and becoming involved.

There is a natural tension between the more career and vocationally directed curriculum offered at the majority of colleges today, particularly in the public colleges and universities, and the more general liberal arts–based programs that emphasize a coherent intellectual experience for students. The latter group aims to provide exposure to the full range of disciplines—the humanities, the sciences, the arts, the social sciences, and languages—that have influenced our development as a democratic nation and people. We will discuss below the important skills for success in virtually all endeavors of work and living. It is noteworthy that the best of the career-oriented programs model the liberal arts curriculum to the extent of requiring students to take courses in the humanistic tradition.

THE SUCCESS OF THE LIBERAL ARTS: WHY DOES SUCH AN EDUCATION WORK?

What is it about a liberal arts education that makes it successful, and what is the definition of success? What are the advantages of the liberal arts "experience"? Here, we will discuss the other factors that make a liberal arts education work.

Positive Effects of the Residential College Experience

It is difficult to extricate liberal arts education from its traditional American home base, the residential college. We are speaking here in general of the classic small to middle-size colleges (1,500–5,000 undergraduate students) that offer a liberal arts curriculum through a four-year program leading to the bachelor of arts (B.A.) degree. And we are talking more specifically about the Hidden Ivies, the select group of colleges we are recommending. They have in common with the Ivy League colleges (see Chapter 2) the advantages of large financial endowments relative to their enrollment, very high admissions selectivity, a top-flight faculty, a large administrative/student personnel support team, excellent facilities for academic and residential life, a strong ethos of community and one's place within it, and very high rates of graduation within four or five years. For graduates, there is a high percentage of placement in graduate schools, fellowships won, eventual completion of masters and doctoral degrees, and support from alumni. In researching the most selective colleges, we have found that particular schools like Dartmouth, Williams, and Princeton had extremely positive ratings by students of faculty, sense of community, school spirit, support of adult mentors and staff, peer interaction and influence, rate of graduation, and graduate school admission.* The positive effects of all of these interacting factors contribute to student happiness, learning, and success.

Factors That Make a Liberal Arts Education Work

These are some of the most important factors that lead to student success during and after a liberal arts education:

•**Faculty commitment to teaching and interacting with students.**
There is a clear emphasis on the value of the undergraduate, the student pursuing the B.A. degree, as opposed to the graduate, the student pursuing the M.A., Ph.D., or other advanced degree. In part, this is because graduate students are found mainly on the campuses of universities rather than colleges. Faculty at the free-standing liberal arts colleges and smaller universities with strongly developed "col-

---

*Inside the Top Colleges.

leges of liberal arts" (or "arts and sciences") tend to self-select. In other words, they choose to serve at these institutions because they are committed to teaching, mentoring, and interacting with their undergraduate students.

•**Opportunities to participate in research projects.** Since there are no or few graduate students present, undergraduate students often have the chance to work with their full-time professors on challenging research, writing, and internship experiences.

•**Intimacy of and opportunity for peer interaction.** Residential living and easy participation in campus organizations—athletic, social, political, journalistic, musical, academic—foster community connections and student-to-student learning and engagement. Since most students live on campus in dormitories, particularly during the early college years, they forge bonds with one another that last a lifetime, and they continue to learn from one another informally during "downtime" and even during formal residentially based mentoring and faculty seminars.

•**Opportunities for leadership.** Due to the smaller size of the colleges and their emphasis on involvement, students can develop their leadership skills in community service, class leadership, sports captaining, club presidencies, and so forth. They can lead a *balanced* life on campus, combining academics, extracurricular involvement, and social life. They can identify themselves in their college community by taking on the mantle of leadership in one area or another.

•**A broader exposure to activities and experiences.** Due to the soft boundaries in small to medium institutions, whether in academic disciplines or extracurricular life, students encounter multiple learning opportunities, pushing them to expand their perspective and to take on new roles.

•**Accessible support services.** Students find it easier to get academic, financial, personal, and career counseling in a more intimate setting.

•**An historic and articulated sense of the college.** This is conveyed through the classroom, the dormitory, and the activities available on campus. Students gain an understanding of the mission of the school, whether through the emphasis on the honor code at Haverford College or a commitment to a core liberal arts curriculum at Reed College. A sense of identity, individually defined by each student but broadly shared by the college community, permeates college life and stays with students when they leave.

•**A strong commitment to developing essential skills.** An emphasis on promoting rigorous critical thinking, analytical abilities, and communication skills, especially writing, inspires the entire curriculum.

•**Synthesis of teaching and research.** Faculty are required to combine teaching with scholarship in order to be up-to-date in their fields and serve as models for students. The teaching and research processes inform one another, as faculty learn from students and combine historical knowledge of their field with current findings and theories.

•**Small classes.** Fewer students in each class and a lower student–full-time-faculty ratio on the whole provide more opportunity for discussion, questions, and collaborative student projects. There is a difference between listening to a lecture twice weekly by a professor in a class of five hundred students, reinforced by one class a week of forty students taught by a graduate teaching assistant (the model at the larger public universities for many of the introductory and even second-level courses), and learning in a class of fifteen to twenty-five, or even forty to seventy-five, taught by a full-time professor in a more discussion-oriented or Socratic format (the model at the smaller liberal arts colleges). In fact, the larger universities have instituted "critical thinking" and "honors" programs, offering more challenging courses taught by professors in a seminar style in order to emulate the format of the smaller colleges. Educators recognize the value of addressing students' multiple learning styles in collaborative, more personalized, and more intellectually stimulating ways.

•**Less emphasis on multiple-choice exams.** With the classroom format and teaching style discussed above, the liberal arts colleges can more easily conduct the examination of student knowledge in ways other than the multiple-choice formats so necessary to test large numbers of students. And when professors give essay examinations or multiple writing assignments, they themselves do the grading rather than assigning the task to a graduate-student assistant. Undergraduates thus receive more informed feedback from professors and are able to express themselves more fully and subtly in their answers. Learning the writing and thinking skills necessary to make coherent arguments is a key facet of the examination process in the liberal arts college.

•**Engagement in the classroom and on campus.** Students have opportunities for engagement on many fronts. The peer influence that raises everyone's personal expectations results in a very high rate of graduation, which varies from 85 to 97 percent of each class. This is extraordinary when you consider that the national rate of graduation from college over a five-year period now averages 40 percent! Even at the most selective large universities the rate of graduation in a normal time frame is only 65 to 75 percent.

Successful Graduate School Admission

As the B.A. degree has become more common in society, the M.A., M.B.A., Ph.D., J.D., M.S.W., M.S., M.D., or other graduate degree has become even more important in distinguishing individual accomplishments and abilities. There is impressive data to reinforce the advantage of choosing the right college on an individual basis, looking for the college that will provide the best foundational and formative educational experience where students can pursue the right courses in their areas of interest. Most families are naive in thinking that one has to attend an Ivy college in order to qualify for the best graduate schools. The liberal arts colleges we review here and many others have equally distinguished records in training and preparing their undergraduates for professional and academic graduate school programs. Let us offer the following as an example in the disciplines of science and engineering for students who attain a Ph.D.: The following Hidden Ivies—Carleton, Grinnell, Haverford, Oberlin, Pomona, Swarthmore, and Reed—on the basis of the number of students per 100 enrolled, send more of their undergraduates on to Ph.D.'s than all of the large and distinguished research universities with the exception of MIT, and at an equal or better ratio than Columbia, Yale, Harvard, and Johns Hopkins. Of course, the latter group's total numbers are much higher. Other liberal arts schools like Bryn Mawr, Amherst, and Williams also have a remarkable record of preparing students for doctoral work in science and technology.

We often talk about attending a top liberal arts college as "not closing any doors." In fact, attending such a college can open many previously unknown doors. Even in an area as competitive and well-known as medicine, liberal arts colleges do an outstanding job of sending their graduates on to become M.D.'s. These students often do not major in a hard science. The Association of American Medical Colleges (AAMC) reports that of all students in 1997–1998 who entered medical school, over 5,000 of a total of over 43,000 students reported majoring in a nonscience subject. For example, 525 history majors entered medical school, with an acceptance rate of 50.7 percent. That compares to a whopping 16,337 biology majors, whose acceptance rate was only 38.1 percent. And 202 philosophy majors were accepted at a rate of 53.0 percent. The AAMC's guide explains some admission goals this way:

> Medicine demands superior personal attributes of its students and practitioners. Integrity and responsibility assume major importance in the classroom and research laboratory as well as in relationships with patients and colleagues. Medical schools also look for *evidence of other traits such as leadership, social maturity, purpose, motivation, initiative, curiosity, common sense, perseverance, and breadth of interests.* Anyone considering a career as a physician must be able to relate to people effectively. The increasing emphasis on a team approach to medical care adds another dimension to the need for this skill. Because of the demanding nature of both the training

for and the practice of medicine, motivation is perhaps the most salient nonintellectual trait sought by most admission committees.*

Sound familiar? Some of these traits sought by medical school admission committees are exactly those fostered by the top liberal arts colleges, where students can pursue coursework and research in the sciences complemented by broad study in other areas of the humanities and social sciences.

Some other reasons for liberal arts colleges' successful preparation for graduate admission are: the close at-hand model of faculty as teachers, researchers, and mentors with a passion for their disciplines; the full use of all campus facilities by undergraduates, since they are not second-class citizens vis-à-vis graduate students and postdoctoral types; the greater opportunity for cross-discipline studies, given the flexibility of faculty and curricula; more funding for internships and fellowships at the undergraduate level; and stronger guidance from teachers concerning graduate schools since they know their students well. We can sum up all of this by noting that the undergraduate is the chief reason for the existence of the college.

Writing in 1987, Marian Salzman and Nancy Better noted the shift in thinking by corporate America toward the liberal arts graduate's skills in adapting, relating, thinking, and "learning how to learn." They offer a wonderful quote from Roger B. Smith, speaking in 1984 as chairman of General Motors:

> There is almost no phase of business life that can be successfully conducted without the benefit of humanistic values and insights . . . [S]tudying the humanities gives you a sense of perspective. In discussions of corporate management nowadays, we often find the word "stewardship"—and with good reason. Certainly, a corporation must serve the interest of its customers, employees, and shareholders. But that alone is not enough. In the cities and towns where its facilities are located, it must also act with a sense of responsibility for the natural environment, the economic health of the country, and ultimately, the welfare of future generations. People who have studied history and philosophy find it easy to maintain such a broad perspective. They know that what a corporation does can have moral implications that may reach far beyond the making of goods and the earning of profits . . . [To] thrive—even to survive—you have to be able to envision new things, as well as new ways of doing old things. You have to extrapolate on the basis of what worked in the past. You have to be able to organize and reorganize your operations to achieve economy and eliminate redundancy. And you need the power to imagine how the course of events might be changed, and by what kinds of intervention.+

*Association of American Medical Colleges. "Medical School Admission Requirements: United States and Canada," 49th edition. 1998 (Washington, DC: AAMC): 22–23. Italics added.

+Salzman, Marian, and Nancy Marx Better. *Wanted: Liberal Arts Graduates* (New York: Anchor Press, 1987): 9–10.

In the information- and service-driven economy of the new century, the skills of the creative learner in utilizing new sources of information in new ways seem even more valuable to employers, public and non-profit agencies, and graduate programs.

Multiple Approaches to Multiple Intelligences

Emotional Intelligence, Emotional Competencies, and Multiple Intelligences are some of the concepts now being discussed as alternative visions of human intelligence. Psychologist Howard Gardner has identified seven distinct types of intelligence, from musical to interpersonal, and continues to refine his conceptual approach to the ways in which we all learn.

HOWARD GARDNER'S SEVEN TYPES OF INTELLIGENCE
WITH RESPECT TO GIFTED/TALENTED CHILDREN

- LINGUISTIC—Children with this kind of intelligence enjoy writing, reading, telling stories, or doing crossword puzzles.

- LOGICAL-MATHEMATICAL—Children with lots of logical intelligence are interested in patterns, categories, and relationships. They are drawn to arithmetic problems, strategy games, and experiments.

- BODILY-KINESTHETIC—These kids process knowledge through bodily sensations. They are often athletic, dancers, or good at crafts such as sewing or woodworking.

- SPATIAL—These children think in images and pictures. They may be fascinated with mazes or jigsaw puzzles or spend free time drawing, building with Legos, or day-dreaming.

- MUSICAL—Musical children are always singing or drumming to themselves. They are usually quite aware of sounds others may miss. These kids are often discriminating listeners.

- INTERPERSONAL—Children who are leaders among their peers, who are good at communicating, and who seem to understand others' feelings and motives possess interpersonal intelligence.

- INTRAPERSONAL—These children may be shy. They are very aware of their own feelings and are self-motivated.

(Source: Southwest Opportunities Program)

Daniel Goleman's study of the particular "soft" skills and talents that make for star performers in the broad spectrum of careers and jobs are, in our opinion, directly related to the experience of a first-rate liberal arts education—in and out of the classroom. Studies indicate that the higher the level of position and responsibility, the less important technical and cognitive skills are and the more important competence in emotional intelligence becomes.

Goleman's definition of emotional intelligence:

> The capacity for recognizing our own feelings and those of others, for motivating ourselves, and for managing emotions well in ourselves and in our relationships. It describes abilities distinct from, but complementary to, academic intelligence, the purely cognitive capacities measured by IQ.*

Goleman's outline of critical competency skills clearly reflects the training in and out of the classroom that a student receives in a challenging, stimulating, diverse liberal arts college.

It takes a high IQ and subsequent grade achievement and test scores to make it into and through college and graduate school today in order to gain entry into the top career tracks. IQ's are generally in a range well above average (115) to superior (125+). What makes the greater differential is emotional intelligence.

Goleman found that employers looking for signs of a person's capability to learn on the job seek the following talents, as opposed to technical skills, in hiring young employees:

- Listening and oral communication skills.

- Adaptability and creative responses to setbacks and obstacles.

- Personal management, confidence, motivation to work toward goals, a sense of wanting to develop one's career and take pride in accomplishments.

- Group and interpersonal effectiveness, cooperativeness and teamwork, skills at negotiating disagreements.

- Effectiveness in the organization, wanting to make a contribution, leadership potential.

As Goleman says, "Of the seven desired traits, just one was academic: competence in reading, writing, and math." In a study of the traits corporations seek in the M.B.A. candidates they wish to hire, the same key skills were identified, the three most sought-after capabilities being communication skills, interpersonal skills, and initiative. Graduate schools of business, in kind,

---

*Goleman, Daniel. *Working with Emotional Intelligence* (New York: Bantam Books, 1998): 317.

look for these characteristics in candidates for admission. Risk-taking, perseverance, independent and creative thinking, and communication skills in all respects are essential traits to develop. Rigid specialization is a road to a dead end in the present and future world.

THE EMOTIONAL COMPETENCE FRAMEWORK

PERSONAL COMPETENCE

These competencies determine how we manage ourselves:

- **Self-Awareness**: Knowing one's internal states, preferences, resources, and intuitions. These include emotional awareness, accurate self-assessment, and self-confidence.

- **Self-Regulation**: Managing one's internal states, impulses, and resources. These include self-control, trustworthiness, conscientiousness, adaptability, and innovation.

- **Motivation**: Emotional tendencies that guide or facilitate reaching goals. These include achievement drive, commitment, initiative, and optimism.

SOCIAL COMPETENCE

These competencies determine how we handle relationships:

- **Empathy**: Awareness of others' feelings, needs, and concerns. These include understanding others, developing others, service orientation, leveraging diversity, and political awareness.

- **Social Skills**: Adeptness at inducing desirable results in others. These include influence, communication, conflict management, leadership, change catalyst, building bonds, collaboration and cooperation, and team capabilities.

(Source: Daniel Goleman. *Working with Emotional Intelligence* (New York: Bantam Books, 1998)

In the next chapter, we review those criteria that we feel are essential in evaluating not only the Hidden Ivies, but any college or university, especially those emphasizing a liberal arts education. Once we have laid out the guidelines for considering and critiquing liberal arts colleges, we will discuss those colleges and universities that we believe do an outstanding job of delivering a liberal education today.

TWO
What Makes an Ivy an Ivy? (And a Hidden Ivy?)

When we talk about "The Ivy League," we are usually grasping at something more than just a name. The designation of being an "Ivy," one of the top colleges in America, carries connotations of prestige, the highest quality of education attainable, intellectual rigor and challenge, and the ultimate in sophistication and selectivity. As we discussed in *Inside the Top Colleges*, it is important for parents and students to look beyond the halo effect that surrounds these colleges and often masks particular academic and social issues that may be of concern. Every college and university deserves similar critical evaluation, but our research and counseling experience has made clear that the Ivies and other most selective colleges share certain qualities that make them truly impressive educational institutions.

Our goal is to discern those characteristics that make an Ivy an Ivy and then to draw together a list of criteria that we have considered in presenting the Hidden Ivies. These are the key factors that families can apply in evaluating any college of potential interest. These criteria serve to showcase the strengths of the Ivies and what makes the Hidden Ivies equal to or almost as good as the Ivy schools. Some of the factors we have identified are quantitative in nature, while others capture more amorphous qualities. *The Hidden Ivies* is not meant to be a rankings guidebook or a quantitative evaluation of these colleges. We aim to provide a set of criteria that help to capture the essence of these excellent institutions and illuminate their particular qualities. We believe these guidelines will be helpful in evaluating any college or university that is striving to provide a top-flight liberal arts education.

THE IVY LEAGUE

What began as an elite athletic conference in the 1950s has grown over time into the collection of the eight educationally elite Ivy League colleges and universities that we know today:

Brown, Columbia, Cornell, Dartmouth, Harvard, the University of Pennsylvania, Princeton, and Yale. What holds them together still is a shared tradition of academic excellence, a high degree of selectivity, and agreement regarding the award of scholarships solely on the basis of financial need. In other words, the Ivies do not offer athletic or merit-based scholarships. In the last few decades, additional colleges and universities have become as selective and as well known as the Ivies. These include Stanford, MIT, Duke, Northwestern, Johns Hopkins, and the University of Chicago. We discuss this group, as well as a few of the most competitive public universities and a few smaller colleges, in *Inside the Top Colleges*.

Every day parents and students in our offices say, "We want the Ivies," or when asked about the colleges in which they are interested, they respond, "The Ivies plus . . ." Often, these families are not even sure which colleges constitute the Ivy League. Nor are they aware of the significant differences among each of the eight Ivy colleges, the level of competition to get into them, and the level of work required to succeed once there. What they do know, however, is the reputation, the wealth of resources, and the academic and intellectual caliber that are available behind the Ivy walls.

What makes an Ivy an Ivy? Important factors in a few key categories are: the faculty; the resources; the educational and leadership team; important educational and other outcomes; the student body; and other special elements that contribute to stature and success. We encourage all families engaged in the college search to use these criteria to evaluate any institution of interest.

THE FACULTY

As Henry Adams said, "A teacher affects eternity; he can never tell where his influence stops." When examining a top-flight academic institution, one is struck first and foremost by the quality of the faculty on campus. And when one talks to students at these colleges and universities, one hears again and again about the importance of the teaching faculty in providing for a good (or bad) educational experience. A number of reports from the Boyer Commission and the Carnegie Foundation for the Advancement of Teaching have leveled much criticism of the 125 or so research universities and the trend toward poor teaching and a lack of concern for undergraduates. It is at the expense of undergraduates that these universities seek excellence and prestige in their graduate programs, and it is the graduate students who do a great deal of teaching at the research universities, freeing up top faculty to do more research and writing. The fame of many of the top research universities is often based on the graduate schools, not on the undergraduate liberal arts colleges. We start with the faculty because they are the living endowment of a college, and it is they who over time serve most strongly to transmit the ideas and content that constitute a first-rate education. What are the important aspects of the faculty to consider?

Academic Preparation

What is the quality and nature of the degrees that the faculty hold? At the top colleges, almost all of the faculty hold the highest possible degree in their discipline. Usually, that means a doctorate (Ph.D.) in their area of academic expertise or, in the case of the arts, an advanced degree in their field of specialization. Additionally, these professors have continued to develop and share knowledge of their field as evidenced in their teaching and writing as well as ongoing training. Look for teaching awards, full-length book publications in academic presses, top journal publications, and the source of undergraduate and graduate degrees as evidence of strong faculty academic preparation.

Accessibility

Good faculty maintain their accessibility to students and a commitment to sharing their knowledge and passion with undergraduates. They inspire and cultivate those in their charge through seminars, after-class meetings, extensive office hours, communications via e-mail, and training in current teaching methodologies.

Work Opportunities with Undergraduates

Look for the opportunities that faculty provide for research projects, assistantships, and internships with themselves or other sources. In addition to classroom work, undergraduates may have the opportunity to pursue publication of joint research projects with their professors, act as research assistants in their areas of interest over the summer or during the semester, and find internships off campus with the assistance and sponsorship of teachers.

Low Faculty—Undergraduate Ratio

What is the ratio of faculty to undergraduates on campus? And what is the percent of time devoted to undergraduate teaching and advising? Low class teaching loads and smaller class sizes allow faculty to devote more time to individual students, to get to know them by name and by their interests, and to advise students on appropriate research projects, courses, and choices of major. Bear in mind that educators self-select, just as students should, the type of academic environment they most value. Those who love to teach and interact with students are more likely to be found in a college that encourages and rewards excellence in teaching.

High Level of Full-Time Faculty

A disturbing trend in higher education has been the demise of full-time faculty members and their replacement by part-time, visiting, or adjunct professors and graduate teaching assistants.

At a lower rate of pay and with less institutional support, these replacement faculty typically have less experience and less commitment to the college. While some may indeed be better teachers than some tenured faculty (those full-time professors who, in essence, after about six years at a university have been offered lifetime employment there), many will move on in a year or two and will not be available to offer advising or support over the long term. The Ivies and other top colleges have a very high level of full-time tenured faculty on campus.

FULL- AND PART-TIME FACULTY

There is a definite trend toward increased use of part-time teachers in colleges and universities in order to cut overhead associated with tenured faculty. It is essential to determine the number and ratio of full-time instructors at the institutions of interest to a student. One of the main criteria for inclusion in *The Hidden Ivies* is the exceptionally high percentage of full-time faculty committed to teaching undergraduates.

PUBLIC UNIVERSITIES	**PRIVATE COLLEGES AND UNIVERSITIES**
Full-time faculty: 45%	Full-time faculty 53%
Part-time faculty: 33%	Part-time faculty: 35%
Research and teaching assistants: 22%	Research and teaching assistants: 12%

A detailed breakout of part-time faculty:

FOUR-YEAR UNIVERSITIES	**FOUR-YEAR COLLEGES**
Public: 21.7%	Public: 21.7%
Private: 31.7%	Private: 43.4%

(Source: Data collected from U.S. Department of Education, National
Center for Education Statistics, 1998)

Endowed Chairs for Teaching Faculty

An endowed chair is a high-paying, tenured teaching position in a specific discipline that is typically named for a generous donor or important college figure. The chair is funded in perpetuity through the college's endowment, a permanent fund for the support of the college. Thus a faculty member holding such a chair has the security of position and freedom from additional pressures to teach a small number of students with a high degree of attention and challenge. As one can imagine, such endowed chairs attract and retain top-flight teacher/scholars.

Faculty Diversity

What is the ratio of male to female teachers and the percent of faculty of color or from non-traditional backgrounds? A diverse faculty contributes to a diverse intellectual environment in which students learn from a variety of perspectives and role models. It helps to foster student diversity, improving the wealth of voices on campus. We believe that such a climate is the most conducive to a balanced and mind-expanding education.

Teacher Training and Recognition Programs

Colleges that care about teaching tend to implement policies and programs that reflect their commitment. Teacher training programs mentor and educate faculty (and graduate students) on the best practices and develop teachers' skills in and out of the classroom. Universities can promote standards to increase the quality of instruction and they can make information available on goals for teaching, student assessments of faculty, and standards for review of teaching practice. In terms of evaluating faculty for promotion and tenure, teaching and advising can be recognized as an important aspect of rewarding professors for their commitment to the undergraduate student.

THE RESOURCES

The resources of a college or university are indicative of its ability to provide a top-quality educational experience for students. From faculty to salaries to computer networks and systems to athletic facilities to library holdings, a college's resource base, best represented by its *endowment*, makes the difference between an excellent and an adequate education.

A Large Endowment Relative to the Size of the Student Body

While there are differences among the individual institutions, all of the Ivies have large endowment funds in relation to the numbers of students (undergraduate and graduate) on campus. One of the best ways of determining a college's ability to provide for its students, faculty, and graduates is to examine the school's endowment in total numbers and as a ratio of the undergraduate population. The total endowment is the overall size of the college's holdings in funds and investments. Income from interest on the endowment allows the college to provide for scholarships and financial aid, teaching chairs, new buildings, maintenance, and residential life. Colleges and universities seem to be incessantly on a mission to raise funds for their endowment, and one can see why. Without these resources, tuition would skyrocket even more than it already has and the colleges would be unable to offer the kind of academic and social experience that students demand.

The commitment of the Ivies and the Hidden Ivies to undergraduate education is in part indicated by the wealth of the school in relation to the number of undergraduates on campus. The ability to provide for undergraduate education means that students entering these top colleges have access to a huge array of resources. Some of the smaller colleges have, relative to their undergraduate student population, a larger endowment than some of the premier, larger state universities. Some key qualities flow from such secure holdings at the Ivies and their counterparts.

COLLEGES WITH LARGEST ENDOWMENTS PER STUDENT BASE

Among private liberal arts universities, Princeton is number three in the nation while Harvard is fifth, even though it holds the largest endowment by far of all universities in the world.

1. Grinnell College
2. Rice University
3. Swarthmore College
4. Emory University
6. Washington and Lee University
7. Wellesley College
9. Washington University in St. Louis
11. Middlebury College
12. Smith College

A number of the Hidden Ivies make the top-25 list in this category, which is pertinent to the factors we include in our discussion of their strengths and priorities. They are:

(Source: National Association of College and Business Officers.
As reported in the *Chronicle of Higher Education*, 1998)

The Quality of the Library

The library (or libraries) is the core of any college. In all its aspects, from computer and information technology to book collections and magazine and journal subscriptions to trained staff and physical plant, the library facilitates a student's ability to learn, research, and study. Technological sophistication allows students to search the college's library, as well as other libraries, perhaps even from the student's own dorm room. Articles and reports in digital form are increasingly becoming essential components of student research papers and assignments.

Ample book collections on campus, particularly in specialized areas, allow students to check out titles quickly and easily, while unique special collections and rare works give students the ability to research primary source material and pursue innovative directions. Prodigious magazine and journal subscriptions allow both faculty and students access to past and current research across multiple disciplines. Some top libraries are even designated holding sites for the Library of Congress.

Strong library resources include professionally trained staff who are able to help students negotiate extensive holdings, conducting sophisticated on-site and computer-based bibliographic and index searches. Such staff can assist faculty in promoting strong research and study skills among students. They often teach seminars in their areas of expertise.

The best colleges have unparalleled modern library *systems*, composed of multiple sites on campus, often separated into specific substantive areas (such as the sciences, the fine arts, English literature, law, and so forth). These libraries have widely available study and reading spaces. They are open in at least one location around the clock, and they are located conveniently to campus. Libraries are an essential learning space for students on campus; it is difficult to overemphasize the importance of their place in one's college career. Ivy-level library systems offer private spheres for student reflection and work, public spaces for the learning community, and extensive resources for the pursuit of knowledge.

Technological Sophistication

Access to the library system is only one important element of campus-wide digital technology. Additional aspects of a technologically sophisticated college community include Internet access throughout academic and residential centers; training opportunities for faculty, staff, and students; a computer or information technology center with professional staff, wide availability of computers throughout campus, and the use of advanced technology for instruction and research. These resources are essential for the most up-to-date use of computer hardware and software, and they provide students with the opportunity to remain at the cutting edge of information usage and analysis. Faculty (and employers) have come to assume a level of knowledge of and facility with computers on the part of their students. They expect advanced research, daily Internet and e-mail communication, papers that are laser-printed and competently formatted on a computer, extensive bibliographies and citations, charts and graphs, and even digitally scanned media. Students should have access to top-of-the-line technology and the support staff to assist them in applying it to their academic pursuits.

Modern Scientific Research Facilities

Up-to-date science and technology facilities offer students and faculty the means to engage in meaningful research. High-quality modern labs with specialized equipment, supported by the necessary research materials, attract and help to retain strong faculty, who in turn lure and work with exciting, high-caliber students.

Residential Life

The quality, nature, and availability of residential housing and facilities for social interaction on campus define an Ivy college. Social facilities can include pubs, dining halls, student centers, and space for student organizations. One must appreciate the magnitude of the impact of a high quality of residential life on the student experience. The informal, lasting, and meaningful interactions fostered by the residential college experience create social bonds and the kind of intellectual community that defines a top college. Colleges are experimenting with "suite"-style living, where a cluster of two or more individual student rooms are joined through a common living room, for example. Special-interest housing, described later, can play an essential social and intellectual role for many students. Some schools are connecting faculty, sometimes on a live-in basis, with particular student dorms of living clusters. And, of course, trained student residence hall advisors play an important role in counseling and mentoring students on an informal basis. Among the Hidden Ivies, almost all the colleges make on-campus residential housing available for 100 percent of the undergraduate student body.

Athletic Facilities

The ability to participate in athletics at many levels across multiple sports allows students additional opportunities to connect with their peers and the college. Top-flight athletic facilities for both intercollegiate and intramural sports, the availability of these resources for all students, the quality of coaches and trainers, equal opportunities for participation for men and women, and a wide variety of athletic programs are emblematic of the Ivies and the Hidden Ivies.

The Arts

Arts facilities, from museums and art studios to performing arts spaces and radio, television, and film studios, are another creative outlet for students. Look for student opportunities to participate in theater, film studies, college radio, and museum research. College holdings may include special art and manuscript collections donated to the school, historic film libraries, specially constructed performance spaces, and individual practice rooms and studios.

Extracurricular Programs

The quality and diversity of extracurricular programs that require college funding goes back to the endowment. Schools with available resources can support a multitude of student clubs and activities, including outdoor clubs, debate teams, community service foundations, highly professional newspapers and journals, internship placements, and political organizations. Extracurricular programs often require physical space, faculty advising, basic funding, and legal and insurance backing from the college. The availability of such programs can indicate an active, involved, and firmly supported student body.

Special Interest Groups

Ivy colleges support and sponsor a variety of special interest groups. In addition to receiving meeting space and financial support, these groups (Asian-American society, African-American student organization, international student alliance, foreign language studies, or college political group) may even have access to so-called "affinity housing." This is an opportunity for like-minded students to live in their own dorm or their own cluster within a dorm. Substance-free dorms, sometimes called "chem-free," coed and single-sex fraternity and sorority houses, and academic honor societies are other living spaces for special interest groups.

Foreign Study and Intercollegiate Consortia Programs

Ivy institutions support an array of special learning centers at home and abroad and programs for international study that are sponsored by the institution itself, often with faculty leadership from the college. These programs allow students to pursue language, cultural, and professional study that is underwritten and facilitated by the college. Additionally, consortia agreements with other institutions allow students to take courses and pursue exchange programs in particular areas of interest. Cooperative 3/2 or 3/3 degree programs allow students the opportunity to pursue the liberal arts at their home institution and then to gain a bachelor's or a master's degree in engineering or business, for example, in cooperation with another institution for an additional two or three years of study. Such opportunities dramatically expand the resources that the specific college can offer on its own.

Support Staff

It is important to note the training and number of staff in support services such as academic, personal, financial, and career advising. Professional resource support for students with learning disabilities and attentional disorders provides assistance in negotiating challenging courses and the entire college system. Career advising helps students to make the transition from school to work and to find internships in their area of interest. Personal, health, and mental health counseling helps students to make positive choices and handle the stresses of college life.

Financial Aid

Our final emphasis in the area of resources is on the financial aid budget of a college. In our opinion, this is one of the most important aspects of a college's resource base and one of the most important indicators of its ability to attract a strong and diverse student body without regard to financial need. This is how these institutions implement their philosophy of a student body based on a meritocracy of talent rather than an aristocracy of wealth and privilege. As families have begun to learn, the "sticker price" of a college's tuition plus room and board, like that of a new automobile, is often not the actual cost that the family will bear in paying for their son's or daughter's education. Prodigious endowment funds for scholarships, loans, and work study allow colleges to defray the financial burden. As college costs continue to rise and students' average levels of debt increase accordingly, this is important. The composition of the aid package offered to a student can vary and reveal major differences among colleges. Some use special talent awards, for example, merit scholars or varsity athletes, to attract and enroll particular students. However, the eight Ivy League institutions have agreed to offer only *need-based* financial aid. Thus they do not fund merit, athletic, or other awards. Yet at the Ivies, anywhere from 45 to 60 percent of students receive some sort of need-based financial aid, and some of these most selective colleges are the only schools in the group of less than ten institutions that are truly *need-blind* in their admissions process, which means they do not take a family's or a student's ability to pay into account *at all* in making an admissions decision.

We believe that a college's ability to provide financial aid for its students and remove as much as possible the money factor in determining which students to accept or reject leads to a more socioeconomically diverse student body composed of bright, talented students from around the world. This diversity helps to make these college campuses interesting and challenging places to study. But even these wealthy colleges will increasingly become subject to the "barbell effect," whereby the neediest students are financially supported and the wealthiest students easily pay the full cost of tuition, but a middle group of students and families, let us say those with incomes between $50,000 and $150,000 (still a lot of money!), will become less and less able to pay annual tuitions of over $30,000 while not qualifying for any aid based on financial need. Thus expensive and competitive institutions could still develop two clusters of students: those with the most need and those with the least need, while those with some need turn to less expensive options like the public universities. Concern over these issues has led a number of the Ivies and other selective colleges to begin reforming their financial aid calculations to make it easier to grant aid to so-called "middle-income" families. However, such moves require ample college funding to be workable, and we are back to the focus on endowment. It remains to be seen whether the Ivies and the Hidden Ivies will agree to give some or more merit-based aid, to take a higher percentage of funds from endowments to significantly lower tuition costs, or to find more creative solutions to these issues.

THE EDUCATIONAL AND LEADERSHIP TEAM

At the helm of any college or university is a team of administrators, faculty, and staff who play an essential role in articulating the mission of the institution, developing its resources, and leading the school through changes in course. Often students and parents do not consider the leaders of a college when evaluating choices in higher education, but we would argue that strong and well-educated leaders foster healthy growth and progress in their institutions. Top colleges are characterized by leaders who often play the part of "public intellectuals," representing their college at home and in the media and taking stands in important current debates on issues. They have the capability to attract to their campus outstanding teachers and administrative leaders, as well as to convince individuals of means to lend financial support to important programs.

Background of Chief Leaders

Focus on the education, training, and professional track record of the president, chancellor, dean of faculty, and dean of academic affairs. These individuals are likely to have advanced academic degrees in a particular discipline; training in administrative, legal, or management areas; and publications in journals, books, and respected magazines. They are members of foundation boards, directors of corporations, advisors to political leaders, and outspoken advocates in public debate. To what degree do these leaders articulate the mission of their institution and their respective ability to carry this out? What positions have they taken on issues of importance to maintaining or changing the core of the school? Look, for example, at President James Wright, recently inaugurated at Dartmouth College. A former provost and dean of the faculty at the college with many years of experience, he has espoused a vision of the quality of campus life and equal access to its resources, and he has emphasized the goal of continuing to develop the combination of top teaching and scholarship among the faculty at this smaller Ivy League college. While in some ways flirting with controversy by taking a clear stand, with the backing of the board of trustees, against predominately single-sex fraternities and sororities, he has begun to lay out a long-range view of how residential life will develop at the school. Effective leaders make statements, set priorities, and bring in talented support staff.

Support Staff

Who are the key people in the academic and student personnel roles and what is their experience? Are they full-time, trained in their areas of expertise? What backgrounds have they had at other institutions? The dean of residential life, dean of admissions, director of the career and graduate counseling center, and the dean of first-year students, for example, can all play significant roles in an undergraduate student's educational experience. These members of the support staff can develop programs, apply for grants, and advocate for students' interests in their areas of emphasis. They can represent and advance their college at professional meetings and through parents and students.

The Mission

What is the mission of the college and how well is it articulated to parents, students, alumni, faculty, donors, and peer academic institutions? How well is that mission implemented? The mission is more than a set of fund-raising goals. It speaks to the college's sense of itself—its history, values, and ideas. The mission is really the identity of a school, and top colleges are clear about who they are, where they are going, and how they are going to get there.

An Honor Code

One of our favorite examples of a distinguishing personality trait of a college is the presence or lack of an honor code. Such a code of academic and personal honesty and principles of good scholarship can be seen as a subtext of campus tone and spirit. It is important to look at the role of campus leaders in expressing the importance of an honor code and to look for examples of its successful implementation. Elements of an honor code—such as the right of students to take exams independently, without professorial monitoring, and the responsibility of students to report any cheating or plagiarism on the part of fellow students—can be reflected in admission application questions that focus on values and ethics. The college may conduct seminars and information sessions exploring the ramifications of the honor code and its place in campus life. Faculty may be required to discuss the honor code at the beginning of each course. Some honor codes are written and require students to sign once or regularly in agreeing to the code's requirements. Others are essentially well-understood and integrated principles that govern campus life. Find a well-articulated honor code and you will typically find a well-developed sense of mission. Here, for example, is a portion of Vanderbilt's statement regarding its honor code and its vital role in the life of the institution:

> The Honor System is a time-honored tradition that began with the first classes at Vanderbilt in 1875. Students established the system and continue to manage it today. It rests on the presumption that all work submitted as part of course requirements is produced by the student, without help from any other source unless credit is given in a manner prescribed by the instructor. Cheating, plagiarizing, or otherwise falsifying results of study are specifically prohibited . . . Responsibility for the preservation of the system falls on the individual student who, by registration, acknowledges the authority of the Honor Council. Students are expected to demand of themselves and their fellow students complete respect for the Honor Code.

While honor codes are certainly about more than just cheating, addressing issues of ethics, social responsibility, and character, they address an all too prevalent issue for today's students. One recent study noted the prevalence of cheating in school, as well as the fact that a "significant minority" of students are interested in addressing the issue of cheating, says Don McCabe, a Rutgers University professor who has been studying the issue for a decade. McCabe is convinced that if those students are given the opportunity to lead the charge against cheating, most others would follow.*

THE STUDENT BODY

The students at top colleges are impressive in many respects. They are scholars, athletes, volunteers, and musicians, with such highly developed résumés that admissions officers, alumni, and parents shake their heads in disbelief, muttering, "I couldn't get in today!" The intellectual environment created by the mixing of a diverse, intelligent, and motivated student body fosters an atmosphere of creative growth, debate, scholarship, and bonding. Look at who the students are on campus and you will gain an understanding of the impact of living and learning in their midst. One of the most important findings in our interviews with thousands of undergrads in the elite colleges was the powerful influence of their peers on the individual's learning experience. Those who expressed their love for the college pointed to the influence of classmates as much as to the faculty.

Selectivity

What does it mean to be a "selective" college? Such a distinction translates into the ability of the school to actively choose among a group of talented students to craft a class comprised of students with unique abilities. The more selective a college, the greater its luxury to pick and choose among the best as it forms each class. Many of the Hidden Ivies have selectivity rates—the percentage of those who apply who are admitted—that approach or equal those in the Ivy League. The selectivity factor plays out on campus in such areas as intellectual preparedness, motivation, range of interests, and diversity of backgrounds, talents, and energy. Such a class fosters the ability of faculty to teach curious and bright students. This attracts dedicated professors to teach at a particular institution and leads to the higher level of intellectual and social stimulation. We know from our conversations with college students and from the research in *Inside the Top Colleges* the major significance of the peer group in defining the nature of the learning experience. Who students study with in the classroom and socialize with in the commons, the clubhouse, and the dormitory is as important to them as who is doing the teaching and the courses they are taking.

---

*Marklein, Mary Beth, "Revealing the answer to cheating" (*USA Today*, 1/5/00): 9D.

Diversity

When we speak of the diversity of the student body, we mean much more than racial or ethnic diversity. This type of diversity is important, yet we add other components for consideration: socioeconomic, geographic, international, intellectual, artistic and athletic, political, religious, sexual orientation. A diverse student body reflects a college's ability to attract, select, and retain students with individual interests and talents, dedications and values, backgrounds and perspectives. Such diversity fosters intellectual and social growth. It forces students to challenge their assumptions and to learn.

Attracting and Supporting the Students

One of the ways top colleges encourage and facilitate diversity is through generous financial aid that is grounded in a significant endowment. We mentioned before how endowment translates into need-blind, or nearly need-blind, admissions at the most selective colleges. These schools have a serious commitment to matching the financial needs of all students. This allows them to attract talented applicants from many backgrounds, to support these students once admitted, and to build a class based primarily on aptitude and interest. Other factors that help to entice and retain top students are a national reputation, high-quality academic and residential facilities, and exciting and well-trained faculty.

OTHER INTANGIBLES

What else is it that drives the engine of a top college? What other key factors seem to define these institutions, together and individually?

Personality and Culture

There is an impalpable dynamic of an overriding culture and ethos that defines a college or university. These elements leave a lasting imprint on a student's education, thinking, and behavior. What is it that creates a friendly or competitive environment, an intellectually curious and stimulating environment, or a hostile and "work for the grades" ethos? Is there a religious or ethical foundation that underlies campus and classroom life or an honor code that establishes the greatest sense of the place? What are the history and the collective traditions of the institution that form its identity and which so strongly affect students?

Single-Sex Institutions

While all of the Ivies are now coeducational learning environments, some of the Hidden Ivies are still single-sex colleges. There is, of course, a great deal of debate on the merits and drawbacks of student learning in such an environment. On the one hand, researchers argue and students maintain, young women excel in a single-sex academic environment, where they are supported to compete in traditionally male disciplines like science and engineering and where socially and intellectually they find an educational setting more attuned to their inclination to cooperate with classmates and learn through discussion. However, others point out that in these kinds of environments, women suffer from a lack of exposure to male cultures and points of view and that eventually women will have to learn to communicate and work with their male counterparts. We should also note that there are a number of all-male colleges that have their own arguments as to why these schools work well for male students. There is a wealth of pedagogic and psychological data in both directions, and what is clear is that the decision to attend a single-sex versus a coed college is largely an individual one. Some young women will do better in one or the other environment, often for very personal reasons. The single-sex women's colleges we have chosen to discuss here do maintain a single-sex setting, but they are linked in one way or another to peer coed environments. They excel academically and offer their students extensive support, understanding of the particular issues that young women face in higher education, sanctuary from the social pressures often found on a coed campus, and an opportunity to learn with and from other high-caliber female students. These single-sex colleges are not for everyone, but they can be important additional considerations for many students.

Prestige

In *Inside the Top Colleges*, we point out the dangers of "the halo effect," of focusing on the name of a college and ignoring the appropriateness of the institution for yourself and the real academic, social, and environmental aspects of campus life. While we would continue to argue against students focusing on names only, it is possible to talk about "the prestige factor" in terms of the institution's reputation and what this prestige confers on its students and graduates. A college's reputation is important, and following the lines of a self-fulfilling prophecy, there is some sort of intangible value to attending a school that is highly regarded. There are the alumni connections to be made, the famous faculty to encounter, the clubby traditions to join and be part of throughout one's life. But prestige is a two-sided coin. You can consider a college's reputation as a real factor in evaluating that school, but you must be careful not to confuse prestige with absolute quality or to choose the "better" school according to myth or rankings without understanding the appropriateness of fit for yourself.

OUTCOMES

In evaluating the "success" of a college, you must at some point begin to quantify outcomes. How well is the college accomplishing what it set out to do? How well is its mission being fulfilled and what measures can you apply to judge those criteria that the school itself sees as important? We have discussed a number of criteria that we feel are important in assessing the quality of any college or university in the liberal arts. Here are some specific criteria that you might use to dig under the surface at these institutions. The Hidden Ivies do very well by these measures:

- The *retention rate* of students at the end of the first and second years. How many students are returning to campus? If this number is low, say below 70 percent, something is wrong.

RETENTION DATA

The American College Testing Program reported their findings that the proportion of undergraduate students who graduate in five years has reached an all-time low. Only 52.8 percent of the students who entered four-year colleges and universities in 1991 had received their degree by June 1996. This represents almost a 5 percent drop in ten years. Additionally, 26.4 percent of all first-year students failed to return to campus in 1998. The retention of first-year students is well over 90 percent for all of the colleges in *The Hidden Ivies*, a record quite similar to the Ivies. The percent who graduate within a five-year period is similarly remarkable, ranging from over 80 percent to a high of 96 percent.

While the national data and rates represent the full spectrum of undergraduate types, i.e., traditional college-going eighteen- to twenty-two-year-olds and more mature learners, they do emphasize the extraordinary rate of retention and graduation of undergraduates at the select colleges that we have included here. We have identified retention as an essential factor in our selection of high-quality institutions because, as we know from our professional experience, the more selective the institution is in its admissions policy, the greater the intellectual motivation of the student body and the greater the sense of belonging to the community. The higher the quality of the faculty teaching and student contact, the greater will be the retention rate and eventual graduation rate. Feeling connected socially and academically are key elements in retention and graduation. Other ACT data show that:

1. 121 highly selective four-year colleges (majority of students graduated in top 10 percent of their class): 8.8 percent dropout rate.

2. 407 selective colleges (majority graduated in top 25 percent of high school class): 18.7 percent dropout rate.

3. 407 less selective colleges (majority graduated in middle to lower half of high school class): 35 percent dropout rate.

These statistics are dramatically different from the thirty colleges in *The Hidden Ivies*, which averaged among the very highest rates of retention and eventual graduation for their entering students.

- Percent who *graduate* in a four- or five-year period. Statistics are now often reported as a five-year graduation rate, which tells you something about expectations of a college career these days. Again, if after five years only 40 percent of students have graduated . . . there is a problem.

- Number who enroll in *graduate schools:* range, variety, and quality of the graduate programs. Today, one's graduate degree, be it in medicine, law, engineering, business, or social work, has become ever more important for job opportunity and security. One of the strengths of the liberal arts colleges is placement in competitive graduate programs.

SUCCESSFUL GRADUATE SCHOOL ADMISSION

There is impressive data to reinforce the importance and advantage of choosing the right college on an individual basis. Most families are naive in their thinking that one has to attend or has an advantage by attending an Ivy college in order to qualify for the best graduate schools.

The colleges we review here, and many others, have equally distinguished records in training and readying their undergraduates for professional and academic graduate school programs. The reasons for this are those we articulate throughout our discussion of the unique qualities these schools possess.

- Level of activity and quality of *recruitment* for seniors, and the overall picture of job placement. If students are not going on to graduate school directly, where are they headed? What kind of corporate and other recruitment is taking place on a college's campus?

- Kinds of *careers* recent graduates have chosen. One would hope to see some diversity in choice here, from finance and banking, law and medicine to social action, teaching, research, and religious careers. However, the kinds of career choices made by graduates can help to indicate the general tone and personality of the campus.

- Number of *fellowships, awards, and honors* received by graduates. Such prestigious awards as the Rhodes and Fulbright scholarships indicate strength of academic preparation.

- *Alumni involvement.* Involvement in annual giving and in such volunteer duties as interviewing, advising on careers, internship mentoring, and job placement are a measure of satisfaction and commitment. Alumni giving represents vast potential in terms of program sponsorship and financial aid and in terms of the operating budget. Alumni networking and communication are a lifelong benefit for the graduate.

SOME LANDMARKS TO CONSIDER IN EVALUATING A COLLEGE OR UNIVERSITY

Statistical Data

Selectivity for admissions, as it represents a quality student body.

Retention of students and graduation rate as a sign of student motivation, loyalty, commitment, and satisfaction, as well as of college resources and programs.

Percentage of students attending graduate school and where they attend, as a sign of motivation and reputation of the college and strength of academic program.

Recipients of graduate and other national/international fellowships.

Job placement and campus visits by companies and other organizations.

Faculty reputation, degrees, accessibility, salary level, etc.

Campus resources for quality of life and educational delivery.

Financial aid programs (percent of students on aid, "need-blind" admissions, etc.), as a sign of ability to attract outstanding students and a diverse student body.

Endowment relative to enrollment.

Student to full-time faculty ratio for undergraduates.

Average undergraduate class size, median class size, percentage of classes below twenty-five students.

Number and breadth of majors and courses offered.

Foreign study and/or internship opportunities and participation.

Participation in sports, intramural, and campus activities.

Overall diversity (racial/ethnic, international, socioeconomic, religious, geographic).

Special Qualities of Campus Life

1. Campus spirit, morale, sense of community, pride in the institution.

2. Satisfaction factor—would students attend again?

3. Satisfaction with fellow students and faculty.

4. A purposeful environment and a balanced social/academic/activity formula.

5. Feeling of academic challenge, academic concern, support.

Applying the Criteria

We have laid out what we intend to be a road map for exploring not only the Ivies and the Hidden Ivies, but also any college or university under consideration. Our hope is that students and families will use the criteria and model of assessment we have presented here to explore any institution of interest. Most data are easily obtainable from guidebooks, news magazines, and the colleges themselves. The factors we discuss represent important elements of a college education for any student, but particularly those pursuing the liberal arts. No institution can be perfect in every way. However, the Hidden Ivies have shown across the board that, according to the criteria that make an Ivy an Ivy, they are some of the best colleges and universities in the country.

THREE
Individual College Descriptions

Following are individual descriptions of the thirty colleges and universities that are the Hidden Ivies. Based on surveys, questionnaires, interviews, and discussions with students and administrators, as well as our own experience with students going to and coming out of these institutions, the descriptions follow a general conceptual format and are presented in alphabetical order. Some statistical data, following our discussion of key criteria in the previous chapter, and qualitative insights gleaned from individuals and published material, capture the individual personalities of each of the institutions.

AMHERST COLLEGE

Wilson Admission Center
Box 2231
Amherst, MA 01002-5000

(413) 542-2328

www.amherst.edu

> Number of undergraduates: 1,650
> Total number of students: 1,650
> Ratio of male to female: 53/47
> Tuition and fees for 1999–2000: $25,169
> % of students receiving financial aid: 44
> % of students graduating within five years: 96
> Endowment: $539,800,000
> Endowment per student: $327,151

Overall Features

Amherst is one of the most illustrious of the liberal arts colleges in terms of its distinguished faculty and highly selective student body. Amherst was founded in the early years of the nineteenth century. Its reputation for academic excellence and its commitment to undergraduate teaching is second to none. Amherst's leaders consider it one of the preeminent residential liberal arts colleges in the nation. While the arbitrary rankings arrived at by the various national publications are considered flawed, nevertheless Amherst is ranked number one among small liberal arts colleges more frequently than any other institution. Its long history as one of the two or three institutions most competitive for admissions seems only to attract more outstanding candidates every year, as well as world-class faculty who want to teach some of the brightest young men and women.

From its roots as a traditional New England college for men only, Amherst has developed into a highly diverse student body of women and men from all corners of the nation and other countries. Amherst is passionate regarding the advantages that the small residential college provides for students and faculty to learn from and with one another. Every undergraduate lives on campus with the exception of fifty seniors who may apply to live off campus. Amherst has long encouraged bonding for life among its students. Alumni, on the whole, speak with passion about their time at the college, of the friends they made, and of their intellectual development. The distinction of their teachers and their ready accessibility is a constant theme of the advantages of an Amherst education. Every course is led by a professor who oversees all of the work done for the class. There are no teaching assistants on the faculty. Even faculty home telephone numbers are listed in the campus directory so that students may contact them. The student-to-faculty ratio is nine to one, and average class size is twenty-two students. Amherst is as selective in its admissions as are the Ivy League colleges, which means that the majority of enrolled students are smart, highly motivated, and well prepared to handle an intense workload.

One tradition that continues is the intense athletic competition with Amherst's oldest of rivals, Williams College. It is considered a successful athletic year if the football team beats the Purple Cows. There are twenty-seven intercollegiate teams for men and women at the NCAA Division III level. Many talented high school athletes are attracted to Amherst for the opportunity to make a varsity team while receiving a first-rate education, especially since the athletic facilities have been upgraded dramatically.

The majority of students join in intramural and club sports. There are some 100 organizations run by students that cover any interest one can imagine. The many music programs, especially the singing groups, are a long and active tradition. A new student center has been well received by students and adds greatly to activities, since fraternities no longer exist on campus. Theme and special-interest residences are sponsored by the college as well.

Since the marked upheavals in higher education in the late 1960s and early 1970s, Amherst has changed as much as any traditional college in a number of ways. Coeducation and a serious campaign to realize greater diversity in the student body have been dramatic for a formerly all-male, virtually all-white, middle- and upper-middle-class population. Students of color made up 29 percent of the most recently enrolled class. Other changes have had a significant impact on the nature of the college as well. The formerly restrictive and highly traditional nature of the curriculum has been transformed into a very flexible, self-directed program of study for each student. There is but one requirement of all students: a first-year seminar. Beyond that, students choose their courses and even have a good deal of flexibility within their major concentrations.

Operating on the assumption that they are among the best and brightest, the academic leaders trust students to choose wisely and challenge themselves intellectually. Amherst students will readily attest to the demands of the faculty and the challenge to keep up with their talented peers. The change in the makeup of the student body is reflected in the curriculum taught today. Courses that focus on gender, racial, political, and cultural themes are many and popular with undergraduates. The rise in financial aid recipients is also a significant indicator of the administration's success in meeting its goal. The once heavily popular fraternities to which almost all men belonged are now a thing of the past. The trustees determined that they defeated the intellectual and social purposes of a small residential community and thus outlawed them from the campus. It is not so easy to block the desires of a smart and independent group of students, however. Even with severe penalties at hand, many have joined underground Greeks off campus. The college has replaced the social role the fraternities once played in campus life with dormitory and cluster social events and intramural sports. Still another significant development since the 1970s is the founding of Hampshire College near Amherst as a highly progressive, student-centered form of education. This then led to the creation of the Five College Consortium, which enables a student to choose courses from the other institutions that include Smith and Mount Holyoke Colleges and the University of Massachusetts. The group sponsors specialized majors and faculty that would otherwise be too costly for any one college to support. Social life and cultural opportunities are expanded for many students as well through the Five College Consortium.

What the College Stands For

Training the minds of young men and women of exceptional talent so that they will think for themselves, understand other people's ideas, write and speak well, and appreciate other cultures are goals the college holds dear. Diversity in all its forms—socioeconomic, ethnic, racial, religious, cultural—is fundamental to Amherst's mission. Here is what the admissions committee states in its literature: "Our commitments to both distinction and inclusion have brought a long line of extraordinarily talented students and scholars to Amherst—people who have enriched our campus, our country and the world. Today, when we select new students, faculty, or staff, we continue to seek talented people from historically underrepresented groups." Amherst concentrates on the liberal arts and on preparing students for lifelong learning and participation in society. As its catalog states, "Whatever the form of academic experi-

ence—lecture course, seminar, conference, studio, laboratory, independent study at various levels—intellectual competence and awareness of problems and methods are the goals of the Amherst program, rather than the direct preparation for a profession."

Curriculum, Academic Life, and Unique Programs of Study

Amherst allows its students great freedom in selecting their courses and designing a major that meets with their interests and academic goals. The single requirement is the first-year seminar, but even here there are twenty topics or themes—all of them interdisciplinary in content—to choose from. There are over 300 courses and 33 majors, with a faculty of 175. Many students create a double major to meet their particular interests. What is called an open curriculum refers to the freedom from set distribution requirements. The academic content of the courses is of a more traditional nature. The key to carrying off this intellectual cornucopia with a rational plan is the faculty and career advisory system, which assists students in planning their academic program over the four years. An awareness of one's intellectual interests and potential future plans plays an important role in designing a plan of study with one's advisor and departmental leader. The faculty encourages students to undertake independent study, research projects, and a senior thesis. Don't forget that all of the departments of the other colleges in the Five College Consortium are also available to Amherst students. About one-third of all students will take advantage of the off-campus and study-abroad programs the college group sponsors. Amherst is also a member of the Twelve College Exchange, which lets a student spend a year at Bowdoin, Connecticut College, Dartmouth, Mt. Holyoke, Smith, Trinity, Vassar, Wellesley, Wesleyan, Wheaton, or Williams. The yearly calendar consists of two semesters and a January interterm. Students are encouraged to utilize this flexible January period to perform community service, undertake an internship with the assistance of the career office and the alumni network, or engage in a project of personal meaning and importance.

Major Admissions Criteria

The selection process is determined by this policy: "We seek qualified applicants from different races, classes and ethnic groups whose several perspectives might contribute significantly to a process of mutual education within and beyond the curriculum." The admissions committee considers the same factors as the other elite colleges in selecting a mere 997 men and women from a pool of 5,200 competitive candidates. High school grades; the quality and rigor of the academic program undertaken; standardized test scores; recommendations from teachers, advisers, or coaches; extracurricular activities and service to the community; essays on the application; and special talents are all considered. "[B]ut no one of these measures is considered determinative. How they intersect makes the difference." This is how an admissions committee can build a new class each year that will reflect the talent and mix of students that meets the goals of the college. A few pertinent statistics will indicate the abilities of those who make it into Amherst: 92 percent ranked

in the first decile of their high school class, the remainder in the second decile; the middle 50 percent had SAT scores of 1,330 to 1,550 combined verbal and math; 20 percent were National Merit semifinalists. Whereas in earlier times a majority of the student body came from a cluster of elite Eastern independent schools and suburban high schools, today some 340 different secondary schools are represented in the entering class. One-third of the places in the entering class are committed to early decision applicants. Since Amherst competes directly with the other Ivies and Hidden Ivies for top students, a highly qualified candidate can help his or her chances by committing to the college with an early decision application. Amherst is able to meet the full financial need of all successful applicants who demonstrate need. Thanks to its large endowment funds, the admissions committee has a need-blind policy in choosing which students to admit. Therefore, if you think you can qualify for admission, be sure not to eliminate Amherst from consideration because of its costs.

The Ideal Student

The typical Amherst undergraduate of the past is no longer dominant in the community. The diversity of the student body and faculty, the freedom to determine one's own academic experience, the absence of the strong fraternity culture, the multitalented cohort of students, and the intensity of the academic program are the key points in determining if Amherst is an appropriate college for you. Those who are capable of gaining admission to Amherst can have many other outstanding choices.

Amherst may be the right college for those who want a very rigorous academic challenge, an opportunity to mix with a wide range of students in a liberal thinking environment, and interaction with faculty of the first level who will push students to perform at their highest capacity. Even with the five college program, students who fit best at Amherst enjoy the closeness of a small college community in a small-town environment.

Student Perspectives on Their Experience

Amherst students feel challenged and supported, focused and expanded by their experiences at the college. Students praise the faculty for its intellectual guidance, advice on securing internships and exploring new areas of study, and willingness to stay involved with students beyond the classroom. They admire each other for their diversity of backgrounds, beliefs, and academic and extracurricular pursuits. Amherst is seen as a supportive environment in which to spend one's college years growing intellectually and socially. For some students, the five college exchange program is a helpful way to expand social, course, major, and resource offerings, while for others the five campuses are far enough apart and the registration process just difficult enough to make the system less workable than would be ideal. Amherst's size and small-town location puts off some students who feel constrained by the fact that news travels fast and everybody knows everybody else. But most appreciate the benefits of close community contact; access to faculty, staff, administrators, and support resources;

and small class and adviser group size. Students love the Amherst graduate network and the bonds they develop with each other and with alums past, present, and future.

What Happens After College

A heavy majority of Amherst graduates continue their studies in the top professional and academic graduate schools. Following the trend of students at the other elite colleges and universities, many graduates will explore different opportunities before enrolling in formal studies. Within five years of graduation, 80 percent of Amherst alumni will enroll in graduate school. History and English, which have fabled reputations due to some of the most distinguished teachers in these departments over many years, ready many students for the premier law and business schools. The sciences also have excellent teachers who provide ample opportunities for research training. A good many students thus gain admission to the major medical colleges or doctoral programs as a result. Here are some statistics the college publishes as indicators of the success of its students in competing for places in graduate schools: 95 percent of applicants are accepted to the top medical schools on the first or second attempt, 84 percent to top law schools on the first try. In the last fifteen years, 120 students have won the most prestigious of the national and international fellowships, which will launch them into the best graduate degree programs. A combination of student talent and the rigor of the education brings such impressive results each year. Amherst fosters a strong alumni network to assist emerging graduates in land-ing jobs or internships of their choice. If any college in our review attests to the value of an undergraduate education in the liberal arts as a foundation for future opportunities in any field of choice, Amherst serves as a standard bearer.*

*A note of explanation to the reader is necessary here. Amherst's administration declined to participate directly in the information-gathering for this book after much thoughtful consideration. There is one reason only for this decision: With due respect for all of the colleges included in *The Hidden Ivies,* Amherst officials feel they belong in the group of elite institutions featured in our other book, *Inside the Top Colleges,* which was completed prior to this work. There is no debate regarding the excellence of Amherst and its place in the top rank of American colleges. However, we had determined to include a representation of outstanding colleges and universities, from small liberal arts to mid-size institutions to large public universities in *Inside the Top Colleges.* Not every elite institution could be included since our purpose was to report on the issues contemporary undergraduates in a variety of selective colleges are faced with in general. In writing *The Hidden Ivies,* we decided not to repeat our work by including any colleges or universities discussed in *Inside the Top Colleges.* Nevertheless, we would be remiss in reporting on the most outstanding of the liberal arts colleges today without including Amherst. The overview of Amherst presented here is based on information the college provides to the public and our knowledge from long-standing contact with students, past and present, and numerous visits to the campus in recent years.

BARNARD COLLEGE

3009 Broadway
New York, NY 10027-6598

(212) 854-2014

www.barnard.columbia.edu

[
Number of undergraduates: 2,200
Total number of students: 2,200
Ratio of male to female: 100% female
Tuition and fees for 1999–2000: $22,316
% of students receiving financial aid: 60
% of students graduating within five years: 87
Endowment: $140,000,000
Endowment per student: $636,636
]

Overall Features

Barnard enjoys a unique situation within the corps of elite women's colleges. While it is a small liberal arts college dedicated to educating talented women for future roles of leadership, it is a distinct component of one of the great research universities, Columbia. As a result of this relationship, Barnard students have a "best of both worlds" opportunity. The college is dedicated to first-class undergraduate teaching, personal engagement with students, and a special concern for the needs of young women. At the same time, the extraordinary resources for learning at a leading research institution are readily available. It would be rare for a Barnard student to complain about limited course offerings, laboratory or library facilities, or extracurricular options. A student can determine for herself to what extent she wants a small college experience or a broader, more diverse learning experience by taking courses of her choosing within the Columbia curriculum. Some students even choose to live in coed Columbia College dormitories. Barnard takes pride in the quality of its faculty, selective student body, and available learning resources. It makes certain that such support services as personal, physical, and academic counseling are tuned in to the needs of young women. Barnard students have the most dynamic city in the world outside the walls of their campus. Virtually every kind of interest can be met. Extracurricular activities are coordinated with Columbia College so that all clubs, committees, athletics, publications, affinity groups, and artistic endeavors are available to all undergraduates on both campuses. Yet Barnard is not every student's cup of tea. Because of its location in an urban setting and its relationship to a very large university, there is less of the intimate residential culture one would find at many of the other selective liberal arts colleges. Barnard's diverse student body and political and academic climate is a natural extension of these factors.

What the College Stands For

Barnard is a standard bearer for the education of women and their preparation for future success in all walks of life, plus an exposure and appreciation of cultural, racial, ethnic, and intellectual diversity. Its student body is composed of one of the largest racial minority populations of any private college in the country: some 35 percent are women of color. The college accomplishes this commitment by virtue of its very generous financial aid program. Financial aid is provided to 60 percent of its undergraduates.

Barnard's president recently wrote to prospective students about the college's philosophy and important attributes: "As a residential liberal arts college, Barnard offers students a faculty of distinguished scholars who remain accessible to undergraduates, along with a dedicated and responsive student services staff. Barnard's New York City setting offers students a world of museums, theatre and music, as well as possibilities for year-round internships in institutions that stand at the center of the fields of commerce, publishing, science, medicine, education, the arts, and finance. As members of one of the four colleges of Columbia University—and the only one to remain independent—students are part of a vibrant 'academic acropolis' on Morningside Heights, which also includes the University's graduate and profes-

sional schools and a number of neighboring institutions, including Teacher's College, the Jewish Theological Seminary, and the Manhattan School of Music. And, as a women's college, Barnard is a place where women never take second place."

Curriculum, Academic Life, and Unique Programs of Study

Barnard's major educational goal, according to a dean, is "to inculcate analytical and critical skills by exposure to a liberal arts curriculum. This exposure puts a premium on breadth in balance with the depth that grows out of the study of a major subject." There are general education distribution requirements in place at Barnard, with a good deal of flexibility in course selection to fulfill the four areas of study. The workload is quite heavy, and since the majority of classes are small, the faculty requires of their students a good deal of writing and independent projects. The humanities and social sciences are the stronger areas of study and the favored disciplines of Barnard women. A good number of premed students add to the mix and they frequently take their science courses in Columbia College. In keeping with its high expectations for its students, Barnard requires that all seniors complete a senior thesis on a focused topic of interest within their major or take a comprehensive examination to demonstrate mastery of the discipline. Another special advantage of the college's location in New York City is reflected in a number of special combined majors. For example, students can achieve a major in music with courses from both Barnard and Juilliard or the Manhattan

School of Music. A student can create any combination of dance, music, theater, visual arts, or writing concentration within and without the Barnard curriculum. A talented young woman can look at Barnard as a school with many options for study, and therefore not have to seek out a more specialized arts environment.

Major Admissions Criteria

In its materials, Barnard notes that "The College seeks women who will benefit most from the Barnard experience: a diverse group of motivated and curious young women who will draw from its deep well of opportunity and contribute to its stimulating community." A dean looks for "smart, self-motivated students, who have a good preparatory education." And, more formally, the college states: "The Committee on Admissions selects young women of proven academic strength who exhibit the potential for further intellectual growth. In addition to their high school records, recommendations, and standardized test scores, the candidates' special abilities and interests are also given careful consideration. While admission is highly selective, no one criterion determines acceptance. Each applicant is considered in terms of her individual qualities of mind and spirit and her potential for successfully completing the course of study at Barnard."

The Ideal Student

A young woman who is truly in search of a number of new challenges will do best at Barnard. The many types of personalities, backgrounds, and points of view that make up the undergraduate student body will stimulate the right person. The college cannot but help take on some of the tone of its environs: There is a competitive, bustling, urban tone exuding a scent of brashness and independence. Barnard women do not hesitate to offer their opinions on issues of importance to them. The workload and quality required of all students can be daunting for someone looking for a less than intense intellectual program of study. With the infinite array of activities and distractions to be found in the Big Apple, Barnard has less of an intimate, cohesive community feel to it. Even so, one dean says that students "should know that Barnard really cares about its students and that faculty, administration, and staff consider them almost like family." The college offers this portrait of its students: "Independent. Intellectual. Interested in ideas. Eclectic. Feisty. Serious. Talkative. Inquisitive. Cultured. Self-assured. Hardworking. Enthusiastic. Politically aware. Adventurous. These are a few of the ways Barnard students collectively describe themselves. Some of these qualities—seriousness, eclecticism, independence—are evident in the experiences and achievements they bring to Barnard. They have graduated from many different kinds of secondary schools and they come from all over the United States and more than forty countries. Their accomplishments range widely. They have been editors of high school newspapers, scientific researchers, published poets, gifted athletes and performers, tutors of disadvantaged children, activists, and committed students. Barnard students are superb resources for one another. They make Barnard an exceptional, exciting institution. But for all of their diversity—ethnic, cultural, social, economic, academic—there is common ground among them. They have great talent and they are open to challenges. They take on new ideas, a rigorous academic program, an international city, and one another with an ease and self-confidence that is distinctly Barnard." Many of Barnard's students spend the greater part of their free time in the canyons of Manhattan. Since most students self-select the type of environment they wish to experience, Barnard women are independent, assertive, creative, mature, and willing to experiment on the whole. There are many other colleges of excellence that would better suit the young woman not yet ready for this stimulating and challenging campus.

Student Perspectives on Their Experience

Why do students choose Barnard? Because of its location in Manhattan, its academic reputation and intellectual caliber, the diversity of its student body, its connection to Columbia University, and, given all of this, its continuing small college feel, which is supported by strong faculty contacts. Says one student, "I chose Barnard because of its easy access to Columbia, because of New York, and because what they say in the brochures is pretty true: Students at Barnard get the best of both worlds. A small college and an academic adviser (unheard of at Columbia) in a huge university and city."

Students appreciate the level of education at Barnard, and are sometimes surprised at the rigor of the academic demands placed upon them. As one graduate notes, "Barnard offers a great education. When I first arrived

here I was hung up on the fact that I was probably better prepared than other first-year students. This may have been true, but my grades my first semester were subpar and this was because I had slacked off. Barnard is not a slacker's college. I think it has a real hangup in that it can be seen as the 'college across the street.' In general, I have had to work much harder in my Barnard classes than in my Columbia ones. It has made me excited about learning and I know I am getting a great education."

Complaints about Barnard tend to focus on a lack of strong sense of school spirit or identity, on lengthy administrative processes, and on facilities. As a student puts it, "In general, I don't view myself as a Barnard student. If you want a pretty much all-girl's environment, it's possible, but I don't, so I do identify more with Columbia. In that sense, I think Barnard doesn't strengthen any sort of school identity. But it must be difficult for any city college to do that. The bureaucracy here is ridiculous; so much waiting in line and traveling from office to office with one question. Also, they seem to be stuck in a time warp: Number-two pencils and line-waiting for registering; the facilities can be dingy because everyone relies on Columbia. Also, Columbia and Barnard offer great speakers and productions and no one ever knows. I don't think they publicize enough. It's like they've given up on a campus feeling."

Barnard students say that those who do best at the college are into the city environment, academics, and their independence. This is not a school for those strongly interested in sports or Greek parties. Those who do well are "motivated," "independent," and "smart," and are comfortable making new friends in what is a new environment for most. Barnard is a fairly competitive place, according to students, and those who enter ought to be prepared to work, to strike out on their own, and to be proactive in meeting new people.

Praise for Barnard centers around its faculty, its academic support and attention to individual students, its more flexible degree requirements, its relationship with Columbia, and its connections to the world of opportunities which is New York. Says a student, "I think it's great that BC [Barnard] has advisors for all their undergrads. All of my friends at Columbia complain about being anonymous and BC kids have someone to address any educational issue to." Barnard is "incredibly diverse . . . The kids aren't dressed similarly . . . There are clubs for basically all groups and ethnicities. . . ." A sophomore says, "It's hard to talk about Barnard College without also talking about Columbia College. They are, if you want them to be, incredibly intertwined. That's what makes it sometimes difficult to identify and be proud of BC. Also, going to school in New York is an amazing opportunity, especially now with it being so much safer and such a trendy spot, but it does have its downside. It's not a typical college experience. But once you go to school here for a year you're hooked. I can't imagine going to school in the country. And it is a lot of work."

What Happens After College

Barnard has a particularly fine reputation and history of sending its students on to top graduate programs. In fact, the college reports that it ranks third among private undergraduate colleges in the total number of students earning Ph.D.'s and in the top five in preparing future female physicians. Some two-thirds of Barnard students graduate having completed internships, many of them with prestigious firms and facilities in their field of interest in New York. These help down the road with career and graduate placement possibilities. Almost one-third of Barnard graduates will go directly on to an advanced degree, most in medicine, law, or business. The college lists the popular employment fields of the remainder as "the arts, communications, teaching, social services, and many other fields."

BATES COLLEGE

23 Campus Avenue
Lewiston, ME 04240

(207) 786-6000

www.bates.edu

[
Number of undergraduates: 1,700
Total number of students: 1,700
Ratio of male to female: 47/53
Comprehensive fee for 1999–2000: $31,400
(includes room and board, tuition, and fees)
% of students receiving financial aid: 54
% of students graduating within five years: 83
Endowment: $145,000,000
Endowment per student: $852,941
]

Overall Features

Its founding in 1855 by Freewill, or Northern Baptists, has left a significant imprint on Bates's social and academic environment. This early religious denomination was in the forefront of the abolitionist movement in the nineteenth century. From its beginning, women and minorities were actively encouraged to matriculate, and the college never had fraternities or sororities. Bates continues its commitment to embrace all voices, points of view, and types of students. Thus for a college of small size located in an old industrial town in rural Maine, Bates attracts a fairly mixed student population. It has become more international and national in recent years, with a record 3,950 applications for 450 places in 1999–2000, enrolling half the class from outside six New England states and maintaining a proportion of the incoming class of 16 percent American students of color or international students. On the whole, students and faculty care about social issues and the environment. This is reflected in the many campus social organizations and various fields of study available to students. Bates has never had to deal with the issue of fraternities and sororities because of its founding philosophy. Consequently, students feel a close sense of community and a concern for their fellow learners, which influences the overall tone of the college. Bates has never had the image of a traditional, preppy, or cliquish college. Faculty are recognized for the quality of their teaching and commitment to inspiring their students. Classes are kept quite small and interactive.

What the College Stands For

Bates is a total undergraduate educational community that focuses on educating motivated, intelligent young men and women in the liberal arts tradition. Its mission is to provide access to individuals of all backgrounds and points of view who wish to learn. It has a quality of openness that is reflective of its mission statement. As with the majority of the colleges in *The Hidden Ivies*, its goal is to prepare responsible young men and women for positions of leadership in their future personal, professional, and civic lives. It emphasizes its concern for social action and protection of the environment through its teaching and programs. Bates has never sought the status of a socially or economically elite institution. A high percentage of students receive financial aid—54 percent—at a college with a smaller endowment than that of the majority of its peer institutions. This is a full measure of its commitment to its purposes. One administrator noted: "For the last ten years, the aid budget at Bates has gone up by between two percent and five percent more than the fees each year, so we are able to meet the full need of all students for four years at Bates, and with sharply reduced loan expectations." Another added: "There is a strong sense of community at Bates, both among our students and between students and faculty and students and administrative staff. There is a collegial spirit of excellence here. Our academic atmosphere drives students to succeed without being consumed by a competitive, cutthroat mentality."

As it approaches its sesquicentennial in 2005, Bates has focused its vision on four principal educational goals, as part of an extensive Goals 2005 planning process:

- Academic rigor and achievement through teaching, learning, and scholarship.

- A distinctive ethos and culture of engagement, civility, and service.

- Connections and an integrating cohesion expressing the promise and value of a liberal education.

- A flexible, principled residential community that is attentive to individual learners.

These goals clearly reflect Bates's two core founding principles of academic rigor and egalitarianism. Bates plans to stay true to its mission and its educational direction in the future. The college requires that all organizations be open to everyone and sees its students as open, friendly, and interested in others. As one dean noted: "Bates is an academic community that is unpretentious and genuine in its interest in the well-being of constituencies within and without. Bates is an involving place—inspiring more students to become engaged in activities outside the classroom that are meaningful to them, add texture to campus life, and make a difference in the lives of others."

Bates has significantly invested in itself over recent decades, tripling its faculty and doubling its student body since the 1960s, adding seven new buildings and seven rebuilt buildings in the last twenty years, and expanding its women's athletic program. Future plans include building a student center, expanding endowment, continuing to try to recruit top students, and bringing more diversity to the college. Bates recently built a Multicultural Center that has already become a "palpable addition" to the campus, improving an established situation of cooperation and communication among diverse groups on campus. Of the faculty, 12 to 15 percent are people of color and 42 percent are women. A dean comments that "technology will challenge us to make the residential nature of the college more meaningful. We need to continue to focus on faculty, students, and facilities (including technological facilities)." Currently, Bates was ranked as one of the five "most wired" liberal arts colleges in the country in 1999 by *Yahoo! Internet Life*. Overall, Bates will continue to focus on emphasizing rising academic achievement levels and the development of the "transformative nature of a liberal education."

Curriculum, Academic Life, and Unique Programs of Study

Bates provides a flexible curriculum with broad distribution requirements. It has a unique academic calendar that is made up of four-four-one terms (four months, four months, and one month). A student can choose a specially designed course of interest, often a non-mainstream topic, for the one-month spring term. Among the over 100 independent study projects completed each year are: alternative education, Horace, ceramics, and bacterial plasmid profiles. Given its mission, Bates's course selection includes a number of multicultural, cross-discipline courses. These include eight new interdisciplinary majors developed in the last ten years: neuroscience, environmental studies, biological chemistry, East Asian stud-

ies, women's studies, African-American studies, American cultural studies, and classics and medieval studies. Classes are small and interactive, and faculty are regarded as available and interested in their teaching. As one administrator noted, "The faculty and staff at Bates come to the college for the same reasons the students do—they get to fire at close range and don't lose people in large lecture halls or between the cracks." It is fairly easy to arrange for a tutorial or special project with a professor in a field of interest, and it is not uncommon for student-faculty research to be published. Bates is well-regarded for the quality of its science programs. A large number of students major in the sciences, especially premedical courses and environmental sciences. The track record of acceptance into professional graduate schools is excellent. Competitive debating on a national level is an active mark of distinction at Bates. Many students are attracted to this demanding endeavor for its rigorous training in logical thinking and communication skills.

Bates prides itself on its service learning program and its international study offerings. With a population of about 65,000, the twin cities of Lewiston-Auburn present communities that have assistance needs that Bates students can address. The Center for Service Learning "uses a collaborative model in their work with social service agencies, involving them as partners in needs assessment and the development of service opportunities for students. As a result, Bates students, through their service, are engaged in reciprocal relationships with those they are working in partnership with." Bates consistently ranks in the top ten among all colleges in terms of student participation in international study. Almost all seniors at Bates write a thesis, a program on which the college puts a great deal of emphasis.

Major Admissions Criteria

Like its neighbor, Bowdoin, Bates makes the SAT an optional requirement for admission. Because of its emphasis on thinking, writing, independent study, and active class participation, the college has found standardized test scores to be a less important predictive factor than a good high school academic record, "extracurricular engagement of substance," and "values," as demonstrated in essays and letters of recommendation.

The admissions office looks for "demonstrated academic achievement in high school, a broad pattern of commitment or interest in others, and particular skills or talents that have been brought to a high level of ability." Bates seeks students who are inquisitive, open, and perhaps with "rough edges." Applications are at an historically high level at Bates, with a 38.4 percent acceptance rate and a 34.9 percent yield rate. Over half of Bates admitted students were in the top 10 percent of their high school class.

The Ideal Student

It is obvious that Bates is not the logical college choice for the urban-oriented student. While Portland, a medium-size city, is within comfortable driving range, Boston is 140 miles to the south and not too easy to reach during the long winter months. For the motivated, smart student in search of a warm, close-knit, socially open environment with an emphasis on outdoor life and Division III and intramural sports, Bates will welcome you with open arms. In the parlance of the contemporary student world, Bates is a very suitable place for an "L.L.Bean" rather than a "J. Crew" type. "Students are apt to be involved in four or five different activities, 'sea anchors' that keep them involved and committed. There are lots of trip wires and support mechanisms: faculty advising, dormitory junior advisors and resident coordinators, mentors, coaches, and the like."

A dean notes that "students who have genuine intellectual curiosity will probably flourish at Bates, and students who have honest and curious interests in others who don't share all of their ideas and background. It's not the right school for someone who wants to sit comfortably in a fraternity or sorority with thirty like-minded people for four years. Bates is not perfect at getting people to move outside their assumptions or backgrounds, but is active in encouraging students to do so." Another adds: "Bates students, for the most part, are modest in their sense of self but clearly recognize that they are a part of a rigorous academic community and have the ability to excel academically, meet the expectations of our first-rate faculty, and positively contribute to campus life."

Student Perspectives on Their Experience

Says one student, "The major reasons that I chose Bates are the friendliness of the student body, the fabulous academics, and Maine!! Bates appreciates and respects the notion of being an individual." Students appreciate

Bates's sense of self and its advocacy of inclusion. They value the academic level of challenge at the college and their access to faculty. "Academics are great, with small classes and cool, really interesting profs. Furthermore, they are challenging and definitely make you reevaluate the way you see the world." Student diversity in its many forms is both appreciated and lamented. "The neat thing about Bates is that you have people with as different post-graduation plans as Harvard Law and working on an organic farm all living and interacting on campus together." Yet "Bates is in Maine, which is not a bastion of worldly culture or ethnic diversity . . . The college is full of interesting and bright kids, and not just a bunch of rich kids either. The kid who lived next to me last year was from Ethiopia and had never seen snow before." A senior notes, however, that "the college strives to bring in as many diverse students as possible and make them feel welcome and comfortable. I have to say Bates is still very white, upper-middle-class though."

While the absence of fraternities and sororities is appreciated, there is still alcohol on campus. A junior student adviser points out that "Bates has numerous activities planned each weekend that do not involve the consumption of alcohol, as well as during the week. There are many clubs and activities worthy of joining." Another negative that students mentioned was the small size of the college, which they felt facilitated "gossip" on campus: "It almost feels like what is your business is going to become, at least, your neighbors'." The flip side of this is that students help each other to succeed: "We all help each other out; without that we would not be able to advance."

Students are divided in terms of feeling they have an impact on campus decision making and would like more communication with the administration, but they expressed the sense that the college, and the faculty in particular, are attentive to their needs. "The college is small enough that we have personal contact with professors and deans, and even the president. If a student wants to get something done, has an idea or whatever, it is possible to get it done. Everyone has a voice, all the time!" A junior resident coordinator notes that "the school has a talented and enthusiastic faculty, with an intense and demanding academic program. It emphasizes the importance of learning to think critically and also of balancing strong academics with a healthy extracurricular activity. The school offers a vast amount of things to be involved with."

Students report that the "well-rounded" student does well at Bates, "a motivated, well-balanced student who knows what it takes to be successful, but also knows how to have fun." Clearly, students should want and appreciate the small college community that Bates offers. "I am impressed with the connection that occurs between students and faculty. The faculty is available to meet whenever needed and is always highly approachable." Another student adds, "Bates makes a family type of community where you can feel comfortable being yourself so that you can meet new people. There is no competition, which makes it even better. The competition is with yourself."

Lewiston does not receive rave reviews, but as one student notes, "I also chose Bates for its location. Yes, some may think this is strange, as the Lewiston-Auburn area is not the most desirable area to dwell in. But I saw incredible opportunities for community service work. I am heavily involved in volunteering with the public school system in looking for ways to boost the aspirations of children in this area." Additionally, students take advantage of the Maine outdoors environment surrounding Lewiston. Overall, "Bates is a great place to spend four years learning and growing. It is a special place." It is special because of "the people and the love and passion they feel for Bates College and the community here, and the support the community offers its students."

What Happens After College

One administrator commented on recent alumni survey findings: "For a college with a long-standing reputation in the sciences, we were surprised to discover we had more graduates whose title was 'entrepreneur,' having founded their own businesses, than graduates who were M.D.'s." Some two-thirds of Bates graduates earn graduate degrees, and there has been an increase in the proportion of graduates who are pursuing business, law, and technology professions, as compared to education, for example. Bates provides "broad preparation for service as well as leadership." A student reports that "the career center is invaluable." It helps students go on to a broad area of jobs and graduate programs. The highest percentage of alumni go on to careers in business and industry, education, health and medical services, and full-time graduate study.

BOWDOIN COLLEGE

5000 College Station
Brunswick, ME 04011-8441

(207) 725-3100

www.bowdoin.edu

Number of undergraduates: 1,600
Total number of students: 1,600
Ratio of male to female: 50/50
Tuition and fees for 1999–2000: $24,955
% of students receiving financial aid: 44
% of students graduating within five years: 88
Endowment: $373,000,000
Endowment per student: $233,125

Overall Features

Founded in 1794 to educate leaders for service to the larger community, Bowdoin has continued its mission in modern terms. It continues to focus on a strong liberal arts education that will prepare students for virtually all fields of endeavor. Its large endowment for a small institution enables it to present a broad range of courses and physical resources to deliver a first-rate education. Its historical emphasis on teaching and interaction between faculty and students in an intimate setting are the overriding features of the college. Its location on the Atlantic coast in a small town in rural Maine impacts the overall environment and, to some extent, the curriculum. "The subtleties and the harshness of Maine, combined with the pride and driven practicality of northern New England set Bowdoin apart from most competitive institutions around the country," says one member of the college community. Students enjoy the out-of-doors lifestyle, with the Bowdoin Outing Club being one of the most popular activities. Students experience an intimate sense of community in a totally residential campus. Faculty are known for caring about students and interacting with them in and out of the classroom. At the heart of Bowdoin is its tradition of academic excellence and participation in a total living-learning environment. Its healthy financial resources allow Bowdoin to attract talented students of all backgrounds irrespective of their ability to pay the costly tuition. Bowdoin is one of the handful of colleges still able to follow a "need-blind" admissions policy.

Diversity and technology development are major themes of the administrative members as they want to prepare the next generation of community and professional leaders. It is expected that the college will grow to just under 2,000, and the faculty will grow as well in order to maintain a ten to one ratio. Bowdoin has the resources to accomplish this, in part due to one of the largest single gifts ever given to a small college: $36 million in 1997. The college hopes to continue its path to becoming more national, rather than regional, in outlook and reputation.

What the College Stands For

Bowdoin recently revised its mission statement, reaffirming old principles and continuing its commitment to them in the future: "The purpose of a Bowdoin education—the mission of the College—is therefore to assist a student to deepen and broaden intellectual capacities that are also attributes of maturity and wisdom: self-knowledge, intellectual honesty, clarity of thought, depth of knowledge, independent capacity to learn, mental courage, self-discipline, tolerance of and interest in differences of cultural beliefs and a willingness to serve the common good and subordinate self to higher goals." Enunciated by the first president of Bowdoin, Joseph McKeen, in his inaugural speech, serving the common good is still present in the minds of all on campus. As one administrator put it: "We expect our students and graduates to serve the common good. We prepare them to do this by teaching them to think for themselves; to learn to speak well publicly; to learn to read critically and write well; to tackle complex research problems and projects; to understand cultures and places other than their home; to develop a sense of history; to develop an appreciation for the arts and music; to understand science and the scientific method; to use infor-

mation technology confidently and to its full potential."

Words that are used frequently by students and staff alike are "leadership," "relationships," and "involved." Students are expected to take the initiative in their learning, to lead their peers and communities, and to give back. Students establish significant relationships with faculty and each other and involve themselves in the myriad activities available on and off campus. The administration and trustees have determined to remove the traditional Greek system from the campus as a means for creating a more unified student body. The class of 2000 contains the last fraternity and sorority members. Similar to other small residential colleges, Bowdoin does not perceive a justification for social units that define and separate the student body. In place of the fraternities, a system of affiliation with a particular dorm or cluster, called the College House System, is being established. "This is a creative and brave step for any institution, and the CH system takes a former strength of the institution and modifies it for today's more inclusive objectives. It continues to give students a sense of place, greater residential options, and opportunities for leadership." First-year students at Bowdoin live in the "Bricks," the traditional first-year dorms, forming the relationships and learning experiences that have been referred to as "the building blocks" of their Bowdoin experience during college and afterward.

A commitment to the faculty and to teaching comes first at Bowdoin. "Faculty are hired and promoted with teaching, as well as research, paramount and the culture of the college fosters strong faculty-student relationships." Students and faculty are said to openly share an enthusiasm for learning, while the atmosphere on campus is described as "open, down to earth, and rarely cynical."

Curriculum, Academic Life, and Unique Programs of Study

"The academic honor code and social code is at the heart of a Bowdoin student's experience. It sets forth a fundamental value of the institution: honesty in all things. We take our students seriously and expect them to approach their education seriously," says a college official. The college does not report GPA or class rank, and "students report that group learning is encouraged and valuable." A large percentage of each class majors in the sciences, particularly the life sciences, as a preparation for medical school or environmental studies. Another significant portion concentrates in the social sciences, such as history and economics, or in languages. The requirements for graduation are relatively flexible in that students must complete courses in each of the four general fields of learning, but they have an extensive range of choices. Bowdoin offers many interdisciplinary majors so there is ample opportunity to create a major concentration around one's personal interests. Such programs include Asian studies, African studies, women's studies, and environmental studies. The college continues its tradition of faculty mentoring students by means of collaborative research in the sciences and liberal arts. Many enthusiastic undergraduates speak of the significance of working on a science project with a knowledgeable teacher or writing a creative novel or a paper on an extensive historical or literary topic. First-year students can choose from more than sixty seminars covering a wide range of topics. Over 60 percent of students engage in one-on-one independent study at some point in their college careers. Over half of Bowdoin seniors complete senior honors projects, involving extensive research, careful writing and revision, and an oral defense. As one administrator puts it: "These projects test many of the things we teach." Another says, "Such courses provide students with wonderful opportunities to test their own wings and to take control of their own education. The independent study promotes intellectual and/or artistic growth, independence of perspective, and confidence." Bowdoin is a member of the Twelve College Exchange Program, which lets students spend a year at any of the other colleges. For some this presents a wonderful opportunity to experience a different setting and community and, in many instances, additional academic studies. Bowdoin students also have ample opportunity to study abroad for a semester or a full year. The college raves about its programs in Asia, South Africa, and Ecuador. "Study Away has an enormous impact on Bowdoin students, providing over half of our graduates with opportunities to expand their horizons beyond the campus. Study Away permits students to test themselves and their intellects in different cultures and institutions and to return more mature, with a wider vision, and almost always a deeper appreciation of Bowdoin and the education it makes available." For international students on campus, the Host Family program provides a home away from home and an introduction to the

United States. The college's newest special program is the Coastal Studies Program, a marine biology and ecology laboratory located twelve miles from the campus. The Outing Club supports Bowdoin's preorientation trips, which help incoming students to appreciate the beauty of the college's surroundings. Bowdoin's Performing Arts Series "brings the world of theater, music, visual, and performing arts to the campus and local community. One of the challenges of rural institutions is to expose students to the world outside the campus (a.k.a., the Bowdoin Bubble). By bringing the arts to the campus, you help expand the bubble and give students a lifelong appreciation of the arts—which is often felt to be a part of the liberal arts vision."

The Student Activity Fee Committee (SAFC) has changed the funding of student activities on campus. An event-by-event funding system has allowed more student organizations to flourish, so that the now more than eighty-five clubs and organizations have more flexibility in programming and funding.

Major Admissions Criteria

Bowdoin turned the tightly knit world of selective liberal arts colleges on its head almost thirty years ago when it made the submission of the SAT optional. The perception of the admissions committee then, and now, is that other factors weigh more significantly in judging qualified candidates. This option has encouraged large numbers of talented and motivated high school students to apply to Bowdoin, allowing the committee to evaluate and accept candidates on more relevant information. "We were the first institution in the

country to institute this policy, and it reflects our value of a concern for the individual and for individuality. We recognize that not all students fit a rough mold of scores, grades, and extracurricular achievement. Over the years, many, many extremely gifted individuals who would not have come had SATs been required, have come to Bowdoin and taken full advantage of this place and gone on to do impressive things in the world (including some of our young trustees)." One staff member recently described the resulting admissions process as a "transcript-heavy selection process." Bowdoin believes that "character should count," looking to recommendations and essays, for example, for evidence of leadership and concern for others.

Teacher recommendations, "particularly as they relate to attitude about learning," student essays, and a student's record of involvement are all particularly important to Bowdoin's selection process. In fact, Bowdoin's supplemental essay requires students to identify the high school teacher who has had the most significant positive impact on their development and to describe that relationship.

Important admissions criteria at Bowdoin are academic ability and potential, "leadership and the demonstrated potential to make a difference in the world," and a "commitment to serving the Common Good." Creativity and "independence of mind" are other factors mentioned. Students are expected to show "excitement about learning, interest in the world of ideas, openness to difference and to new areas of knowledge, courage and a willingness to grow and change, the desire to be part of a residential community."

Bowdoin is seeking to increase the diversity of the college, even as it reports 18 to 20 percent students of color from all over the world, including America. Utilizing its need-blind admissions and resources, the college continues to try to expand socioeconomic diversity, particularly in terms of maintaining its historic mission to provide a top education to citizens of Maine, the poorest New England state, whose students represent the second highest proportion of students on campus by state origin. Secretary of Defense William Cohen and former Senator George Mitchell, now brokering peace in Ireland, both attended Bowdoin as scholarship students from Maine in the 1950s and early 1960s.

The Ideal Student

The qualities that Bowdoin looks for, and that are said to help students succeed, are self-motivation, high academic ability, leadership in multiple arenas, social conscience, and "the ability to think for oneself, not simply to do well on a multiple-choice exam, but to be able to write, compose, and create an original work." Students are curious, active, and independent. They are not competitive with each other, but rather supportive of one another's success and learning. They are problem solvers, critical thinkers, adventurous, and confident. One admissions officer noted: "They are incredibly adept at juggling numerous activities, and they are rarely spectators."

As one dean put it: "Maine is known for its physical beauty and the friendliness and rugged individualism of its people. Bowdoin reflects that. When you enter Maine on Interstate Ninety-five, the sign says 'Maine:

the way life should be.' That applies to Bowdoin, too: the way college should be." Bowdoin takes pride in its location and sees it as a major strength. Students should be looking for a personal, friendly campus, nestled in a rugged outdoor environment. "Some families may feel that Bowdoin is too isolated, but I contend that around this coastal location, which is full of natural, accessible beauty, one can develop and profit from exposure to a sense of wonder. It is hard to put into words, but Bowdoin and Maine are magical places because of their proximity to the sea, and this very unusual coastline, which is rocky and inhabited by mast pines and all sorts of unusual creatures."

A long-serving academic dean observed the ways in which Bowdoin students grow during their education at the college: "I see students who have found their own voices. Engagement and confidence, shaped by some sense of humility about the unknown, are key indicators of the personal and intellectual growth of Bowdoin students. All of these qualities were exemplified, for example, by the five environmental studies coordinate majors I met at a lunch yesterday where they made presentations about their summer internships. The clarity and enthusiasm of their talks, the thoughtful and careful ways that they answered questions, their confidence in taking on challenging tasks with great responsibility, and the breadth of their perspectives both on their particular tasks and the larger enterprises in which they are engaged helped to exemplify for me the success of our teaching/learning community at Bowdoin."

From the admissions office comes this point: "Bowdoin is not for everyone. If a student desires an urban setting, if he or she is driven by grades or is content to be a spectator, Bowdoin would not be a good college choice. On the other hand, if he or she is unabashedly excited about learning, is not only unafraid of new challenges but embraces them, Bowdoin may be a good fit."

Student Perspectives on Their Experience

Students show evident pride in the college, its personal qualities, the relationships they have developed, and the access to faculty that they enjoy. They are somewhat more concerned over a perceived lack of racial, ethnic, and income diversity on campus, although this is not a universal feeling. What comes through most clearly is their sense of Bowdoin's excellent teaching, familial environment, and beautiful location. "What Bowdoin does best is to truly let the students value and appreciate our teachers. They play a very active role in the students' lives not only in the classroom, but outside the classroom as well," says one member of the class of 2001. Another student offers: "The school is tremendously strong academically. Its size also allows students to develop mentor-type relationships with faculty, staff, administrators, trustees, and alums. These relationships complement the academic education by offering a personal education about life. These relationships, which I couldn't have imagined before attending, are perhaps even more valuable than the formal education I've received (which is very valuable in its own right)." About the classroom, we hear this: "On average, class size is about sixteen, which is very comfortable. This maximizes class discus-

sion, and classes are less about taking notes, more about intellectual discourse." An international student says: "The relationship with professors has been extremely rewarding for me. Several of my professors have become more than professors to me. I consider some of them my friends. Bowdoin has given me more than education: I believe it has changed my personality in a positive way."

Students succeed at Bowdoin, says one junior, because of "the laid-back academic atmosphere and the student support. Yes, the work is hard and challenging, but we don't let it get to us, we don't stress out. We don't let the work control us. In addition, there is no competition among students. You don't have to be afraid to tell people you got a C on a test, it's OK to share notes. Students here help other students with work as much as possible. It's not a battle for grades here, it's a joint effort to learn!" Students caution that "the only people who have a hard time at Bowdoin are shy and unwilling to get involved right away in class discussions and extracurricular activities. These things are critical at Bowdoin for both social and academic reasons." A recent graduate adds: "Bowdoin promotes individual thought which reaches beyond a set of lectures and a textbook to make learning a reformation and enrichment of thought processes. It allows the opportunity for extracurricular activity, which is pivotal to a well-rounded, academically enhanced, and generally happy education."

Academically, students reflect the value of the broad liberal arts education at Bowdoin, especially the flexibility but also the extent of the distribution requirements. "One of the best traits about my education at Bowdoin has been the diversity of fields that I've taken courses from. While I'm an economics major, roughly half my courses are in the humanities. I don't think I would have had that opportunity at a larger university." Students value their academic opportunities and their faculty, and, if they are critical at all, it is about the need to promote more diversity on campus, to open the administration to student concerns, and to try to weed out from the campus those students who do not take their education seriously or appreciate the opportunity they have been given to attend Bowdoin.

Students we heard from clearly chose Bowdoin in part because of its physical environment. "Bowdoin could possibly be in the best place for any college to be. First of all, we are nestled amidst the beautiful Bowdoin Pines. The town of Brunswick is a terrific small town with restaurants, bars, theaters, stores, banks, mini-malls—everything you could want in a college town, most of the things within walking distance. Freeport, one of the best shopping towns on the East Coast, is only fifteen minutes away, while a major city, Portland, is only twenty-five minutes away. Bowdoin is also minutes away from the coast and within an hour or two you could take advantage of the great outdoors, be it skiing, hiking, climbing, rafting, etc. Even though there is an abundance of things to do on campus, the number of things to do off campus is endless."

Students leave Bowdoin with a sense of the place and their connection to it and each other. One student going on to teaching says: "Part of teaching is also learning, and I'm sure that with a liberal arts education, I will be ready to teach in almost any subject within reason. I like the challenge of being able to shape minds and challenge them. I had a rough time as a teenager, and my teachers were among the few people who I adored, so I hope to have that kind of impact on my own students." Another says: "Bowdoin is special in many ways. The community feel at Bowdoin is unparalleled. You know the checker at the dining hall, the lady at the mailroom, and the security officers on a first-name basis. Professors are there to help educate, not to lecture for an hour then return to research before passing out poor grades. You are a person at Bowdoin, not a statistic. You are a face and a person who contributes and learns from the entire Bowdoin community."

Another student tried to capture the essence of the college's history and its impact on her: "I have yet to find the words to describe the feeling that I get when I am here. I knew from the very first time I saw Bowdoin that this was the place for me. Maybe it has something to do with the history of the college. Each incoming freshman is invited into the president's office to sign 'the book' that the likes of Nathaniel Hawthorne, Henry Wadsworth Longfellow, General Joshua Chamberlain, former Senator George Mitchell, and Olympic gold-medalist Joan Benoit have signed during their first few days at Bowdoin. It is an inspiring feeling, to sit in a classroom where Harriet Beecher Stowe sat, or to look up at the same huge oak trees that Nathaniel Hawthorne saw when he stepped out of his classroom. The students at Bowdoin will always aspire to greatness, and maybe that's what makes it so special. It is an indescribable feeling of honor to be a Bowdoin student; it is the past, present, and future students of this college that make it so special."

What Happens After College

An inordinate number of Bowdoin graduates pursue a career in public service, following their sense to serve the common good.

About 75 percent of Bowdoin graduates enter graduate or professional school within five years of graduation, from Ph.D. programs in the sciences to medical colleges to economics. Many Bowdoin students go into teaching at the secondary school level. Other areas include the film industry, the business world, advertising, law, and volunteer programs like the Peace Corps and Americorps.

With the personalized attention that Bowdoin students receive and their opportunities for extracurricular and co-curricular study, projects, involvements, and internships, they develop their potential at the college. "A Bowdoin education," says one administrator, "enables students to develop many of the critical abilities needed to succeed in today's workplace: problem-solving, analytical reasoning, written communication, public speaking, and leadership skills." Bowdoin alumni are "fiercely loyal" and willing "to bend over backward in helping students to make connections through networking; locate internships, summer jobs, and full-time positions; as well as relocate to different parts of the country. When you come to Bowdoin as a student, you become part of a very large extended family for the rest of your life. And wherever you may travel in the world, you will run into other Bowdoin people, and when you do, you will be bound by a common thread that is somewhat inexplicable to others."

BRYN MAWR COLLEGE

101 North Merion Avenue
Bryn Mawr, PA 19010-2899

(610) 526-5152

www.brynmawr.edu

Number of undergraduates: 1,200
Total number of students: 1,900
Ratio of male to female: 100% female
Tuition and fees for 1999–2000: $23,360
% of students receiving financial aid: 55
% of students graduating within five years: 77
Endowment per student: $204,210

Overall Features

While Bryn Mawr is no longer officially affiliated with the Society of Friends who founded the college in 1885, the philosophy of the Quakers is instilled in the community, as witnessed by the diversity of the student body, the high percentage of scholarship recipients, the emphasis on intellectual exploration and the freedom to voice one's opinion at all times, and the honor code. Bryn Mawr was one of the early colleges founded to educate women in an academically and intellectually rigorous curriculum equal to that which men received at the Ivy League universities. Bryn Mawr puts a great deal of emphasis on its small classroom, teacher-led approach to educating young women. As one senior administrator points out: "The college remains passionately committed to the proposition that women can do and be anything they choose, a commitment that origi-nated in the vision of founding dean and second president M. Carey Thomas that Bryn Mawr would offer women the best education available." Bryn Mawr prides itself on its sense of an intimate community with strong faculty-student interaction and opportunities to explore virtually all fields of interest despite its small size. Enjoying a sizable endowment for a small college and with an open relationship with its neighbor, Haverford College, Bryn Mawr enables its students to experiment with myriad academic and extracurricular programs. If a course of a more exotic nature is sought, a young scholar can take advantage of the college's relationships with nearby Swarthmore, Villanova, and the University of Pennsylvania. Bryn Mawr is a picturesque suburban campus located only fifteen minutes from Philadelphia. The honor code that permeates both the academic and social life of the college is a key element in defining the environment.

What the College Stands For

A senior administrator provides an excellent summary of Bryn Mawr's educational goals: "Bryn Mawr endeavors to encourage and develop a student's ability to think creatively, critically, independently; to express herself articulately; to experience the excitement and joy of learning; and to acquire the flexibility of intellect to continue to learn and adapt throughout her life." A dean notes of the college's expectations for students' learning that "in the end, one hopes that they have learned how to seek answers for themselves, as well as the pleasures and satisfaction which come in finding them. One also hopes that they learn, both in the residence halls and classrooms, on the stage and in the gym, a great deal about negotiating differences of value and opinion in productive ways, and about working cooperatively." The college tries to promote a diverse and international student body, with about 10 percent of students coming from abroad and about 25 percent being American women of color. Bryn Mawr is small, only about half the size of the other women's colleges in the Hidden Ivies (Barnard, Mount Holyoke, Smith, and Wellesley), and sees itself as having "a strong sense of student community and a very accessible and engaged faculty, deans, coaches and other student life administrators." Students are helped to succeed by, as one dean notes, "a supportive environment in which the students have close contact with the faculty who are enthusiastic and supportive and in which students are encouraged to work collaboratively with other students."

The college's future priorities include increasing its emphasis on international stud-ies and interdisciplinary programs, promoting increased faculty diversity (17 percent currently are persons of color), and developing the use of technology and computer science in instruction, research, and linkages to neighboring colleges. The college will seek to make more internships available to students and build on its current strengths in undergraduate research. Bryn Mawr expects to increase its academic competitiveness and has just hired a new director of athletics. And the college "has an excellent arts faculty in areas ranging from dance to creative writing to theater, with a strong music program available to Bryn Mawr students through Haverford College. We expect to support students' increasing interest in both curricular and co-curricular opportunities in the arts."

Bryn Mawr's library holds over 1 million volumes, quite a large number for any undergraduate college. The college sees its links to Swarthmore and Haverford as being integral to its mission and availability of resources. "The college has long been renowned for the rigor of the education it offers, and the culture of learning on campus means that it is possible to learn as much if not more from conversations and discussions with one's peers as one learns from the faculty." Students from all four classes—freshmen through seniors—are mixed in Bryn Mawr's dormitories, encouraging diversity and communication in the college community. A final word from one dean captures the college well: "Bryn Mawr is a very open, congenial, supportive, collaborative, and diverse institution, which at the same time is rigorous with many longstanding traditions. The mix of old and new, small (small college) and large (large urban environment), American and international together with a student body and faculty from a wide variety of backgrounds makes Bryn Mawr an extremely interesting and exciting place."

Curriculum, Academic Life, and Unique Programs of Study

Students feel the honor code helps to cut down on the academic competition and create a real sense of intimacy among the community members. The tone is one of working hard in a challenging program and deriving satisfaction from a job well done. This is an important factor as Bryn Mawr insists on completion of a fairly strong set of distribution requirements in three divisions: the natural sciences, social sciences, and humanities. First-year students take a seminar of choice focusing on analytical and writing skills. All students must complete the equivalent of three years of foreign language studies. Bryn Mawr has been long known for outstanding science programs and faculty, and one-third of the students major in one of the sciences. As one administrator notes, "Bryn Mawr is being recognized increasingly as the most successful college in the United States in encouraging women to pursue further study and careers in the basic sciences." There are ample opportunities to work side-by-side with a teacher in a research project that provides an advantage for graduate school admission and success. "By learning in a purposely small institution, students are able to test themselves both in the classroom and in a culturally diverse community, to learn both broadly and in depth, and to ultimately take a leadership role in their field,"

says one administrator. A significantly high proportion of graduates go on to medical school or doctoral programs. Here, too, students are encouraged to work with their faculty mentors in research projects and to do independent study. Bryn Mawr also has excellent departments in the humanities, such as English, art history, languages, classics, and archaeology. It ranks first in the nation in the percentage of graduates completing doctoral degrees in the humanities. In all fields, "the curriculum relies on hands-on research, primary sources, and few or no textbooks." Although graduate students generally do not teach courses, "the tradition of small graduate programs in many of the disciplines makes the academic setting more complex and interesting," states one academic dean.

In addition to the student-administered honor code, the Bryn Mawr Self-Government Association—the first such association for students—"reinforces the independence and integrity" of the students. "There is no dean's list, no Phi Beta Kappa, no published grades, even though Bryn Mawr is one of the most selective colleges in the United States. Every student establishes her own standard of excellence and the measures of success are one's own," says an admissions official. The college encourages study abroad, community service, and research or career-oriented internships and externships. Students are supported by a network of health services, a system of "peer tutorials," and class deans. Each student is assigned to an assistant dean for academic advising, and "the faculty and dean's office work closely in providing a seamless program that promotes academic persistence and campus adjustment," according to one administrator. The collaboration with Haverford, only about a mile away, is the closest of that with any other neighboring college and is accessible and well-utilized in terms of students taking classes and even majoring at the Haverford campus. Additionally, the college participates with Haverford and Swarthmore in the Tri-College Summer Institute for entering students of color to help make the college more accessible. The college continues this effort with an Affirmative Action Compliance Plan for faculty searches, a full-time admissions assistant director to coordinate minority student recruitment, an Office for Institutional Diversity with a full-time director, and multiple clubs and cultural organizations on campus.

Major Admissions Criteria

The Bryn Mawr admissions offices ask questions in three main areas in their evaluation of applicants:

- Has the student challenged herself in the most academically demanding program available to her? Has her performance distinguished her from her peers?

- Has the student made the best use of the resources available to her? Has she been recognized for exceptional academic promise and do her teachers and school recognize her intellectual creativity?

- Has she made contributions outside the classroom?

Thus the college focuses primarily on a student's high school curriculum and performance, personal qualities (such as "independence, responsibility, integrity, maturity") as shown in interviews, essays, and recommendations, and evidence of key talents and interests.

The Ideal Student

There is a set of characteristics that defines the successful and satisfied Bryn Mawr woman: focused on her education, intellectually curious, intelligent, motivated to achieve, and open-minded. One dean refers to the college as a "community of serious student scholars." An admissions officer notes that the Bryn Mawr Book Award is given to "an outstanding female student in the junior class who demonstrates a love of learning and a curiosity about the world around her." He comments, "Bryn Mawr students are generally those who have distinguished themselves in their secondary school and stand apart from the crowd in talents, interest, and academic seriousness. The most successful students are those who are unafraid to tap into the array of resources available to them." Another dean describes the following qualities that help students to succeed at Bryn Mawr: "intelligence, intellectual curiosity, independence, willingness to work hard, ability to manage one's time and to balance serious academic commitment with engagement in the community, and openness to and curiosity about other people from very different cultures and backgrounds." With its commitment to embrace all ideas and points of view, together with its socially diverse community, the college encourages open dialogue on political, intellectual, cultural, and gender issues. This is not an environment for the faint-of-heart student who is more comfortable with a conservatively inclined, homogeneous population. The social life revolves around campus activities, lots of sports at the intercollegiate and intramural levels, and time with one's friends and studies. The bright young woman who is looking for a large, socially driven, traditional college environment will not take well to Bryn Mawr.

Student Perspectives on Their Experience

Students work hard at Bryn Mawr, and they appreciate the college's intellectual focus. The college "makes it OK to be smart, to work hard, to stay up night after night reworking your thesis. You're not dorky, not nerdy—you're a Mawrtyr. There is tremendous potential in a women's college. Bryn Mawr exploits that potential in a wonderful way—women (of course) do everything here!" Another student says, "Bryn Mawr is best at giving its students confidence in their abilities and helping them learn to respect themselves and the other students around them." Students feel supported as female scholars, and also "know how to have fun and enjoy ourselves while we're working hard." Successful students are "open-minded, hard-working, and dedicated to succeeding," and are involved academically and socially. "Bryn Mawr is full of multitalented, fun, and intelligent women."

Traditions at Bryn Mawr are felt to be very important. "BMC's traditions really give the students a sense of being part of a group and belonging at BMC. The honor code has also been really important to me because it gives me the freedom to be myself academically and socially, because I know that my peers respect me and will treat me honestly and honorably." A junior says: "There are awesome traditions, beginning with Parade Night, where the classes get to know each other at the first Step Sing of the year, moving on to Lantern Night, where the frosh choose sophomore 'hellers' for one week as initiation into Bryn Mawr, and finally, May Day, where the year wraps up with celebration, music, and fun. The traditions really bring students together. Each class has their own set of songs that they sing at Step Sings."

Academics are important to the students: "If you work hard and love learning, you will pass. You have to be good to get good grades, really good. You have to be on top of things, hardcore in to what you study. But that's how we are." Deans help students to succeed, and older students help the younger classes. "That's why the mixed-classes dorms are wonderful. The older students have taken the classes, dealt with the profs. They can help oh so much." Continuing the theme of student cooperation, another student says successful Bryn Mawr students are "people who are willing to work hard, but not just to compete with other students. People who are interested in learning for the sake of their own knowledge and aren't interested in grade-grubbing. The honor code really helps students to focus on challenging their own personal limits and not worrying about how they compare to other students." Some students, however, note that there is a "facade" of a "noncompetitive and low-stress atmosphere," while "this place is a pressure-cooker ... Students are fiercely competitive and driven. Bryn Mawr is intense." Students are advised to "learn how to study and what works best for you ... The secret of success is to study what you love. Find something you are passionate about. Liberal arts offers so many options."

There are some calls to make the linkage with Swarthmore more usable, to improve athletics, and to push students to get involved in more off-campus programs. Perhaps one senior sums up the campus life best: "BMC is a school/community. It is difficult to be flat or one-sided or to hide. It is demanding academically but in a good way that makes learning more concentrated and bonds very intense. It is getting through something together while opening the world to individual goals at the same time. The word that comes to mind is 'balance.' BMC teaches balance while also supplying it. There are pressures and there is support."

According to one student, "Lesbians have a very strong voice on campus. And it makes sense, doesn't it?" Students do recognize the diversity on campus, but see the college as moving in the right direction, meaning that there is work to be done. Within the diversity, one student sees most students with similar personalities "conservatively minded, politically liberal, studious, and introverted. Students not of this type tend to hate Bryn Mawr and transfer out." The Customs program, however, is seen as extremely important by students in helping them to make the transition from high school to college.

The faculty are strongly valued by students. Says one, "Since I have been at Bryn Mawr, there is something that has helped me

a lot, and that is the following: The profs are just people. They are not here only for me, but they are here to help. They mess up sometimes, too. They burn toast on occasion. Talk to them. See what kind of people they are. Profs can help you only as much as you let them (really) and don't be too defensive—they are really smart, you know!" Another states, "Professors are incredibly supportive and helpful for each student in the class." And, "You really get to know your professors. It's not uncommon to eat over at your professor's house or baby-sit their children."

"I would have to say that what Bryn Mawr really does best," says a junior, "is make your college years memorable—with beautiful traditions, amazing academics, and lasting friendships." Says another, "Bryn Mawr is a unique and special place ... it offers the best of everything—academics, traditions, small social functions like coffee houses and then bigger parties at Haverford, the adventure and excitement of a big city, the ability to meet lots of new people from so many places." Women are empowered at Bryn Mawr, that is for sure. Says one graduating senior, "As a senior I have been thinking about this a lot and in May I won't feel like a flower that's opened. That's high school. This May I will feel much stronger, like a firecracker—everything at Bryn Mawr is intensified."

What Happens After College

An administrator says, "In addition to their commitment to their academic work and the vibrancy and vitality of the campus community, Bryn Mawr students are passionate about improving the world. As the school that educated the first woman to win the Nobel Peace Prize, Emily Balch, the college has attracted and gained from the presence of students whose ambitions, beyond their own successful careers, have included living lives that contribute to the social and common good." Education, law, and medicine attract many alumnae, but so do business, management consulting, and international careers. The majority of Bryn Mawr graduates will earn graduate degrees. The percentage of students who go on to complete a Ph.D. is and has been among the highest in the country. As one student and future veterinarian/Ph.D. says, "I want to show the world that Bryn Mawr women are smart, strong, and can do anything they put their minds to."

CARLETON COLLEGE

100 South College Street
Northfield, MN 55057-4016

(800) 995-2275/(507) 646-4190

www.carleton.edu

[
Number of undergraduates: 1,860
Total number of students: 1,880
Ratio of male to female: 47/53
Tuition and fees for 1999–2000: $23,469
% of students receiving financial aid: 60
% of students graduating within five years: 89
Endowment: $462,000,000
Endowment per student: $245,745
]

Overall Features

Carleton was founded in 1866 to provide for outstanding students in its region of the country a liberal arts education comparable to the very best of the New England colleges. To this day the college maintains its focus on training young men and women in the intellectual skills and moral and ethical values that a first-rate liberal arts education provides. It has continued and expanded its commitment to a wide mix of talented students of all backgrounds. It enjoys a significant endowment for a college of its size, which it puts to good use in providing financial aid for 60 percent of its students. Carleton also offers a large number of national merit scholarships out of its own funds that can cover all or most of the tuition. It can truly state that it seeks talented students who can gain admission irrespective of ability to pay. Carleton is a socially, politically, and intellec-

tually liberal environment. It stimulates students to think critically and open their minds to many points of view. Students are attracted to the college for its reputation for academic excellence, a major commitment by the faculty to teaching and mentoring students, its spirited and close community, and its success in graduate school admission.

Faculty members are generally known for excellence in teaching and knowledge of their disciplines. Many have achieved renown through research and publishing. The student body is serious about both studies and active participation in extracurricular organizations. While its location is known for the frigid weather that prevails, the tenor of the community is one of warmth and friendly engagement among students and faculty. Reflecting the tradition of training students for future intellectual and leadership roles, Carleton graduates have won an unusually high number of

Rhodes Scholarships over the years. Over one-half of any graduating class goes on to leading graduate schools. In fact, Carleton ranks number one in the nation among liberal arts colleges in the number of graduates who attain doctoral degrees in the sciences. It ranks among the top five in doctoral recipients in all disciplines. There is no Greek system in place, so the social emphasis is on dorm and college organizations for social and leadership life. An unusually large percentage of students get involved in community service organizations in the larger community. Carleton has a strong tradition of providing the best campus facilities possible. To this end it recently completed a $10,000,000 recreational center. While Carleton is a beautiful campus with a total residential population in a small town, the twin cities of Minneapolis and St. Paul are only forty miles away. This allows many outlets for social fun and community service. Athletics are active but not intensely competitive.

What the College Stands For

Carleton's president, Stephen Lewis, succinctly describes the college's mission in a letter sent each summer to parents of incoming students: "A Carleton education is first and foremost an experience that helps young people learn to think clearly, analyze critically, and express themselves effectively. And, as a residential, liberal arts college, Carleton has always been concerned with the development of the entire person. We offer a rich smorgasbord of activities beyond the classroom to give students many opportunities to develop all their talents and sensibili-

ties, and the kinds of students who come to Carleton take advantage of them." Another administrator describes Carleton's goals as the following: "To assist students in their intellectual and emotional growth and to encourage their physical and spiritual health by providing a wide array of activities and support; to equip students with the academic and personal skills to become independent, successful adults; and to encourage students to lead balanced lives."

Carleton sees itself as a place where issues are taken seriously, but people at the college, including students, do not take themselves too seriously. There seems to be a sense of humor to the place, as illustrated by the student tradition of displaying a bust of Friedrich Schiller, which was "liberated" from the president's office in the 1950s, at major events. The bust is kept in secret by a group of students and occasionally wrestled away by other students during a showing. It drops in by helicopter, is paddled down the river in canoes, and is ushered into auditoriums by ninja-clad students! A Carleton administrator describes the college's uniqueness by the degree to which faculty are involved with students outside the classroom; there is support and shared activity among the students themselves; competition of students against one another is absent from the culture; there is physical activity and life outside the classroom relative to the hard work within it; friendliness and decency are present; myths about the weather have an impact upon the ethos of the campus; and there is a sense of humor in an otherwise serious-minded campus population."

Future initiatives include a continuing

effort to promote racial diversity, curricular development within and among academic disciplines, and several key building projects. The college will have a great deal of support from alumni for these initiatives: With alumni fund participation at over 50 percent, alumni gifts last year totaled $18.8 million, up from $2.8 million a decade ago.

Curriculum, Academic Life, and Unique Programs of Study

The distribution requirements necessary to graduate are both flexible and rigorous in nature. There is a wide choice of courses to fulfill the goal of the liberal arts education, but every student must also complete the equivalent of four years of a foreign language. Students are required to complete their education in twelve terms, and there is a four-year residency requirement. A senior project that demonstrates a comprehensive understanding of one's academic concentration is required. Many interdisciplinary and major/minors are available to students, and the college's music programs are seen as particularly strong. Many take advantage of the outstanding science departments and collaborative research with faculty. The recent completion of a $100,000,000 capital campaign has led to the building of a number of new buildings, such as the Center for Math and Computing, the Biology Center, and the Olin Hall of Science. Two-thirds of Carleton students seize the opportunity to study abroad for a semester or a year. It is the rare young scholar who has not been able to build a particular field of study around his or her interests with the support and encouragement of the faculty.

Although people choose to enroll in Carleton for its academic excellence, there is not an intense competitive quality to the community. Collaboration in learning, support for one another, and individualism in the intellectual enterprise are encouraged.

Carleton puts a great deal of emphasis on faculty development, "hewing teachers out of Ph.D.'s", and sees a "symbiotic relationship between students and faculty." There is faculty review via student letters; there is a senior faculty mentoring program, a Learning and Teaching Center for teacher training, and new faculty workshops make for a fine continuum of training and support for faculty. The benefits of this emphasis for Carleton students are clear. Students are helped to succeed by "small classes and an outstanding commitment on the part of faculty; strong academic support programs, especially in modern languages and math; the students themselves—a deeply caring, very friendly, and supportive group of people; a tradition of intellectual engagement; a student life staff committed to helping students learn how to live in a complex, diverse world; and literally hundreds of activities to choose from."

Students have a voice in the college through the College Council, which includes five elected students, five elected faculty, two staff, a trustee and an alumni observer, and the president. This is a voting body that impacts college policy and represents the participatory nature of Carleton. A student senate controls a $300,000 budget for student organizations. Eighty percent of students work at least ten hours per week on campus, and some 30 to 40 percent are active as volunteers at any given time.

One interesting new project stems from a $5-million grant from the Star Foundation to support a comparative Asianist program and faculty position. This will be a cross-cultural program and interdisciplinary minor, where ten students each from Asia and the United States will work together. Carleton currently teaches Chinese, Japanese, and Russian, and sees this comparative program as a key component in developing its strength in this area. The grant will also help to provide financial aid for international students. Carleton has a 16 percent minority student population, with an undergraduate grant and research program for minorities. There are over a dozen multicultural student organizations, an Office of Multicultural Affairs, a Summer Math and Science Program for minorities, and other support programs on campus.

Major Admissions Criteria

Applications, and yield, are up at Carleton. In 1999, applications increased by more than 20 percent to the largest number ever—3,457—with the acceptance rate falling to 46 percent of applicants. Carleton ranks number one among four-year liberal arts colleges in the number of National Merit Scholars in the freshman class, seventy-four. A student's academic record and intellectual interest are clearly important in admissions, as are involvement in school and community and character traits. Of incoming students in 1999, 60 percent played one or more varsity sports in high school and 60 percent participated in one or more musical groups. Over 82 percent were involved in significant community service. Carleton enrollees are "lively, friendly, and active."

The Ideal Student

Individuals who love learning in an intimate and interactive academic environment will thrive at Carleton. This is not the right environment for the student who prefers to remain anonymous in the classroom or who does not like to engage in active learning. Studies come first, but a high premium is put on community involvement and service. Those who care about social problems and taking action will find a host of opportunities and an encouraging environment at Carleton, and so will those who like to participate in leadership activities or low-key athletics at the intercollegiate or intramural level. Given its diverse student body and generally liberal-minded tone, a candidate for Carleton should be open to new people, new ideas, active debate, and discourse. One should have concern for social issues and should be willing to work hard at one's studies.

At the same time, Carleton students are not seen as being competitive with each other, or elitist. As the president of the college said at a recent talk for new students, "As part of our Minnesota roots, Carleton people tend to be pretty self-effacing—lots of Garrison Keillor 'shy people.' You'll find very few who are self-promoters, and that distinguishes us from a lot of other places. This is a diverse community of strongly individualistic people . . . A typical Carleton student is one who would be outraged by the notion that there was a typical Carleton student!" Students who are "freer spirits" and "fierce individualists," "curious, active, seekers and risk-takers" will feel at home at Carleton.

An administrator notes that "the campus

culture has moved significantly over the past decade. There is less insistence on independence and autonomy for their own sakes, and more acceptance of the responsibility that accompanies freedom and independence. Students are more open to guidance—but not interference—in their personal lives and their groups. Students have continued keen interests in engaging in multiple activities, physical recreation, and intensive study, and seem very open to working in groups."

Student Perspectives on Their Experience

One student tells a revealing story about how he chose Carleton. "In selecting a college to attend, I worked from larger concerns to more specific school decisions. After realizing that a liberal arts experience could only be had during undergraduate education, I decided to attend a small school where I, rather than the literary journal, would be the focus of a professor's attention. Once this was determined, I asked a high school guidance counselor for the best liberal arts school in the country. She said the best learning environment was at a place called Carleton College, so I applied, was accepted, and arrived for new-student week six months later." He went on to become student body president.

This student and others find the open, involved learning environment highly rewarding. "At its best, Carleton fosters an environment where students are truly free to examine differing ideologies and confront new ideas under the guidance of an involved, student-driven course. But this is done outside the classroom as well, with significant intellectual challenges arising anywhere from the library to the dining hall to the weight room. Also, I recall going over to my professor's home with my classmates, who numbered seven total, and discussing the implications of the cold war on current U.S.–China policy over dinner and after-dinner treats." Students find help at the Write Place for editing papers and at the Math Skills Center.

A 1999 graduate says, "Carleton is an exceptional school for students who do not know what career path they want to take immediately after college, but who are highly self-motivated, like to explore many different opportunities, and are smart and focused on academic excellence. The school provides what I would say is an ideal environment for this type of student to develop to his or her full potential (insofar as a college education can help)." Nevertheless, "while students at Carleton are encouraged to push academic boundaries, the school is not a good environment for those who are not certain of their political, social, or religious beliefs. The 'liberal' sociopolitical environment is not helpful to those who are still exploring their social, political, or religious beliefs, regardless of whether or not they end up agreeing with the prevailing college culture. Especially at risk are students from non-mainstream backgrounds (racial minorities and the like) who are not firmly committed to getting a good academic education."

"Students who are smart, study hard, keep goals in mind while maintaining a flexible and explorative mind, are self-motivated, willing to take the initiative to make use of college resources, explore new things with the guidance of professors and fellow students, and who are able to keep in mind that the world is a lot larger than a small college campus in rural Minnesota do best at Carleton." Students are said to benefit greatly from informal faculty friends and mentors, but one student says, "The formal academic advising system is a shambles, and students are really on their own. The corollary to being free to explore your options is that the college really can't step in to help out until it's too late." A junior notes, however, that "the school excels in academics and maintaining high standards of excellence not only with students, but also with professors and faculty. There really isn't anyone at Carleton who doesn't belong there, and that goes for both students and faculty. The school does best in making its students succeed in every aspect and seeing that nobody falls through the cracks. You have to try very hard to be a failure at Carleton." The professors "are truly amazing. Not only are they brilliant, they are so anxious to talk to students about their subject that students can't help but become excited about their studies."

"There isn't a way to construct a quintessential Carleton student and the same goes for who does the best. Everyone who goes through Carleton finds something unique about himself or herself," says one student. Another adds, "Carleton is a place where students can experiment both academically and socially, in a beautiful setting with exciting, intriguing people."

A note on the environment: "A really warm coat makes walking across campus to go to class in minus-fifty-degree weather a whole lot easier." And one will need to go out, since the students who do best are "people who take advantage of everything! There are so many clubs, concerts, talks, and events going on every single day, that those who don't go are really missing out." And "who else has a 'Pie Day' on March 14, late-night trivia, ice sculptures, concerts, plays, events, and so many truly fun opportunities. There are a lot of outgoing, creative people here."

Complaints focus on alcohol use and abuse, with requests for the college to take a more active role in prevention, and there is the wish to improve multicultural recruitment, retention, and involvement. "Even having one multicultural student drop out is devastating to our already small community, let alone half a dozen. It seems to me that there aren't enough multicultural faculty and that the college isn't trying terribly hard to overcome this problem." On the other hand, a junior says, "Some of the requirements Carleton imposes that are supposed to ensure diversity overdo it a bit ... Carleton *loves* diversity. While we don't have many racial minorities, we do have a wide range of students from different socioeconomic positions and just a lot of really neat, different, interesting people." And a senior, while arguing for improving the diversity on campus, adds, "I hope as we try to increase our diversity we don't do it for numbers sake, but for enriching the lives and educational experiences of young people."

One student sums up his experience by saying, "Carleton, in my opinion, provides the best, most challenging undergraduate education in the world. It is not a practical education, but students graduate with the ability to do pretty much anything. This is because of one thing: While Carleton does not have anywhere near the resources of the more traditional Ivies, the compensation is that Carleton professors love what they do, love to teach, and are darn good at it. Carleton's faculty are consistently ranked some of the best in the nation." Says another, "Even if I didn't want to succeed, Carleton would make me succeed and the faculty would, and does, personally see to it that I do just that." And another maintains, "I couldn't have picked a better college. I love the fact that I can walk across campus at eleven P.M. after a Model UN meeting to find friends studying in Sayles, people in the science and computer labs, and students even playing Frisbee or hockey long after dark. The energy around Carleton is invigorating and I adore it."

What Happens After College

One administrator notes, "Alumni continually comment that Carleton gave them the confidence to try many different things and, in fact, that is what happens. There is no typical career path, so alumni scatter around the world in business, law, medicine, education, research, government, community service, religious vocations, journalism, and politics." Carleton has an extensive career center to serve students: "Special programs and workshops teach students how to write résumés, interview successfully, select a major, search for jobs and internships, and many other valuable skills. Students may tap in to a vast and supportive alumni network to conduct informational interviews and arrange career-shadowing experiences." Some 17 percent of Carleton graduates are enrolled in advanced degree programs six months after graduation, including academic study, law, medicine, and other areas. Six months after graduation, 66 percent of Carleton graduates are in the workforce. The largest numbers are in business-related fields and education. Service programs, such as the Peace Corps and Americorps, are also popular. "A Carleton education prepares students to excel in many different fields. Students learn excellent communication and analytical skills, and have ample opportunities to develop their leadership skills. Students are also encouraged to develop a commitment to their communities, diversity, and global awareness."

CLAREMONT-MCKENNA COLLEGE

890 Columbia Avenue
Claremont, CA 91711-6425

(909) 621-8088

www.mckenna.edu

Number of undergraduates: 950
Total number of students: 950
Ratio of male to female: 57/43
Tuition and fees for 1999–2000: $20,760
% of students receiving financial aid: 59
% of students graduating within five years: 86
Endowment: $321,000,000
Endowment per student: $337,894

Overall Features

From a chronological perspective, Claremont-McKenna is a youngster among a select cohort of liberal arts colleges founded one to two centuries earlier. Nevertheless, it has come of age at a rapid pace due to its quality of faculty, facilities, focused curriculum, and setting. Its sizable endowment is remarkable for so young an institution. The commitment to assisting talented students to attend the college is made possible by this endowment. A small college as measured by its enrollment, it presents the course offerings of a much larger institution. Claremont-McKenna was one of the five founding colleges that comprise the Claremont group. The others are Pomona, Pitzer, Harvey Mudd, and Scripps Colleges. Each has its own unique style, campus tone, and emphasis in educating its students, but in combination they provide a marvelous range of learning options for the serious student. All classes at the five colleges are available to all students. There does not appear to be much red tape involved in registering for courses outside one's home institution.

Claremont-McKenna's academic content, from the start, has been centered on government, economics, management, international relations, and public policy. As one dean describes it, it is a liberal arts college on the crease between the academic and real worlds. Students are attracted to Claremont-McKenna for the opportunity to prepare for future roles in the business, consulting, legal, and governmental arenas in a strong teaching-oriented program. The college attracts top faculty in these particular disciplines who are interested in inspiring bright students to master their subjects. The student-to-faculty ratio is quite impressive and thus classes are small, allowing for a good deal of interaction.

Students are expected to speak up and speak out in an intelligent and articulate fashion. A great many students will undertake independent projects and assist faculty in their research throughout their four years. With an eye on their future in national and global leadership positions, a significant number will study abroad. The college environment has a traditional and pragmatic tone, given its particular academic emphases and its students' professional ambitions. There are very few liberal arts colleges where one-half of the student body are registered Republicans. Claremont-McKenna encourages exploration on and off campus, and leadership experience is available through the many clubs and organizations. There are no fraternities or sororities at Claremont-McKenna, so social life is centered in the dormitories and in campus organizations. For so small a college there is a very active athletic program, with about one-third of students involved. Students join with Harvey Mudd and Scripps Colleges for Division III intercollegiate competition. The college is located only forty miles from Los Angeles, so there is ample opportunity to experience a large urban center.

What the College Stands For

Claremont's mission statement emphasizes the college's major goals: "Its mission within the mutually supportive framework of the Claremont Colleges is to educate its students for thoughtful and productive lives and responsible leadership in business, government and the professions and to support faculty and student scholarship that contributes to intellectual vitality and the understanding of public policy issues." With an average class size of nineteen and about 93 percent of students living on campus, Claremont carries out its mission in the midst of an intimate residential college setting. "We receive incredible support from trustees, alumni, parents, and faculty. Our tight-knit community provides an environment in which students may learn and grow as future leaders," says one administrator. Another institutional leader notes: "Claremont-McKenna College provides an education that is both highly pragmatic and deeply philosophical: We want our students to learn how to think independently, creatively, and analytically; to learn how to utilize their thinking skills and apply them to real-world matters; and to possess both the ability to recognize opportunities and the courage to take risks."

The college sees itself as providing "an environment in which it is easy to maintain a high level of morale—regardless of the fact that the academic demands are intense. CMC students genuinely enjoy learning—in classes and from one another. Student life is campus-focused, with virtually all students living on campus and freshmen, sophomores, juniors, and seniors all living in the same dorms." The college emphasizes that "teamwork at CMC is a given," and that "support for CMC students is everywhere." One dean notes, "This is a first-rate faculty dedicated to undergraduate education," which helps the students succeed. A well-known historian, Jonathan Petropoulos, who was recently hired by Claremont, spoke clearly of his and the college's commitment to both teaching and research when he was quoted as saying, "This is the paragon of a liberal arts college. It's an atmosphere ideal for learning, yet it has the resources of a research university."

Curriculum, Academic Life, and Unique Programs of Study

Given its academic emphases, Claremont-McKenna can best be described as a traditional model of education, what one senior administrator called a "focused liberal arts college . . . I believe we are better defined and focused than most liberal arts colleges." Both faculty and students tend to put their energy and thought into the subject matter at hand, which means learning the nature and dynamics of social, economic, political, global, and managerial material. The distribution requirements are traditional as a consequence of the curriculum. Students must gain exposure to the full range of arts and sciences with more specific courses than those required at a majority of Claremont's peer institutions. The curriculum consists of one-third core and distribution requirements, one-third major requirements, and one-third full electives. All students must complete an in-depth senior thesis in their field in order to graduate. With its collegiate neighbors, Claremont-McKenna provides students with fifty different study-abroad programs to choose from. A majority of the student body will take advantage of this opportunity at some point in their four years. For the student who determines that he wants to attend medical school or other science programs, the excellent level of science courses and facilities at several of the other Claremont colleges are at hand. Another unique feature is the presence on the same campus of nine research institutes, all of which are focused on various areas of professional leadership, public policy, and the law. Ambitious undergraduates can avail themselves of their presence to gain research and internship experience. In sum, Claremont-McKenna holds a special position within the world of small liberal arts colleges. It presents the same range of liberal arts curricula as other top colleges but with a concentration in the applied social sciences and a much wider course selection thanks to its four affiliated colleges and its graduate management program and the research institutes.

Other programs of note include the Athenaeum Speakers program, which brings well-known "historians, authors, poets, politicians, musicians, professors, economists and more" to campus four evenings a week; a 3/2 Management/Engineering program, in which students complete three years at CMC and two years at a school of engineering; the Robert A. Day 4+1 M.B.A. program, in which economics and economics/accounting majors can earn a B.A. from CMC and an M.B.A. from the Drucker Graduate School of Management at the Claremont Graduate University across the street; the Washington, D.C., semester; and the Practicum Program. In terms of student support, an administrator points out that "faculty advisers offer guidance in course selection, career paths, majors, and general life issues. Student sponsors help students during freshman year with everything from class selection to adjustment to college life. And the career services department helps immensely in providing internship and job opportunity information and guidance."

We can expect CMC to stay true to its

founding mission and identity in the future. As the college welcomed only its fourth president in 1999, it emphasized the recruitment of "superlative" students and faculty and technology. Says one administrator, "We will seek to build our name recognition on the East Coast and seize an unassailable position on the top-ten list of private U.S. liberal arts colleges." A dean says "'Steady as she goes.' CMC's mission is still fresh although some adjustments will occur," notably in increasing commitment to and enrollment in the sciences and modern languages.

Major Admissions Criteria

Claremont looks for students who are "intellectually able, ambitious, and self-disciplined." One administrator lists the most important admissions criteria as "high school grades, SAT score, and record of activities." Another says, "CMC attracts students who take the initiative, whether in a student leadership position, as captain of an athletic team, or president of a club. We want students to be involved and to stand out as leaders." The major admissions criteria are:

1. High school transcript—challenging courses along with high grade point average.
2. Extracurricular activities— demonstrated leadership.
3. Scores—SAT or ACT.

A dean notes a particular interest in students' participation in student government.

The Ideal Student

It should be obvious that Claremont-McKenna provides a particular brand of education in a unique environment in comparison to the majority of liberal arts colleges. The student who is passionate about the social sciences, management studies, and leadership training in the public, private, or international fields will find this college most attractive. Conservative in nature, Claremont-McKenna best serves those who are either conservatives themselves or who welcome the opportunity to challenge themselves in such a setting. The workload is demanding but there is an emphasis on collaboration and support among its student body and with the faculty. An intelligent young man or woman who might be considering an Ivy League institution with a vision of preparing for a career in politics, government, law, or business should seriously consider the educational foundation to be acquired at Claremont-McKenna within a more personalized environment. Here is one college where a young man can bring his blazer, white shirt, and tan bucks.

In this environment, however, the student body is relatively diverse, with about 5 percent international students, 5 percent African-American students, and, reflecting its California environment, about 13 percent Latino students and 22 percent Asian-Americans. There are a number of organizations on campus to work with diverse populations, and CMC seeks to promote diversity by offering substantial financial aid packages. Overall, however, "CMC students are themselves focused, ambitious, and driven. They're also friendly and open to working with other students in groups, and eager to support each other's achievement and success," says an administrator. Students who do well are "ambitious, self-confident, intellectually curious, and socially outgoing." They are "positive and upbeat with a can-do attitude. CMC challenges and pushes students to the limit, but at the same time, students are given the tools and resources to succeed. CMC combines thought-provoking classes with a warm, family-type environment."

Student Perspectives on Their Experience

"I think that our best attribute as a college," says one sophomore, "is the competitiveness to be the best and to achieve excellence. It is dynamic, and always trying to progress . . . It is also a great networking opportunity. CMC puts its students first and takes care of us."

The students who do best are "those who are the most driven to do the best. There is a heavy workload, so those who can organize their time and stay focused are the most successful." Small class sizes and "a great faculty" help. "The whole environment is conducive to a family-like atmosphere where students and faculty help one another. With small classes, you can't get lost in the shuffle. Also, having all of the finest and latest resources (technological and others) is greatly appreciated." Within this environment "the freedom that the school gives you to conduct yourself as an adult without lots of personal restrictions helps students to grow as people. And the willingness to help out in every way possible with any activities one wishes to plan makes living here enjoyable." A senior notes, "It's very helpful to be able to reach our professors at home during weekends and around finals

time." And "anyone who puts in enough time and effort is able to perform well at CMC. No class is extremely difficult if the student is eager enough to visit the profs and/or tutors."

Students succeed "who really like to get involved (academically and socially). Those who take their classroom learning beyond the academics (with internships, research, on-campus organizations) find themselves intellectually stimulated and *busy*!" Echoing this, a sophomore says, "Really smart students with no personality have no place at this school. The school trains leaders, not mid-level management. If you have no voice, no opinion, and don't get involved, you probably won't get a whole lot from CMC." Students help each other to do well. "There's a strong sense of solidarity among our student body (particularly when it comes to surviving classes). We help each other out a lot." Another says, "Takers do best at CMC—students who know what they want and take it. Vocal, forward, strong-willed, self-confident students succeed here; it is not a school for the passive." This sentiment is echoed by other students, both positively and negatively.

A senior campus leader planning to go into informational technology consulting says, "The management-engineering program offered me the opportunity to meet a lot of the science department faculty, who are important to my physics major interest." Another notes the attraction of the Claremont system: "The five college program is really neat. I get to live on a government-, economics-, and foreign policy–based campus (one that, because it has concentrated its funds in the above list, has no music department) and still can be a music major by taking those classes at Pomona

College. The system also tremendously changes social life; because all the campuses are adjoining, I essentially go to school with over six thousand people. But I share my deans and administration at CMC with only one thousand. I have the community bits of a small school with the social options of a larger university."

Some students complain about wanting to make the "study material a little more culturally diverse," although "the other Claremont Colleges make up for that." One quips that the college "trains corporate attack dogs quite well" and that "it lacks the typical leftist slant, and thus has little truck with ethnic studies or feminism. Whether it studies these fields in a 'worse' manner than others is not necessarily bad." To illustrate the contrast, another student says, "Strong support of the ROTC [Reserve Officer Training Corps] program has been important to me. Support like that goes against the grain of increasing 'politically correct' and overly liberalized schools." General campus diversity, however, is seen as positive, particularly when including the other Claremont campuses. Students would like to see CMC more well-known and publicized on the East Coast, and a visit is clearly crucial in evaluating the college; potential students have given their campus visits glowing reviews. Some students do not see every area of the college administration (i.e., the registrar) as following through on the small school model where all students are treated as individuals. But "student government has an incredible budget and is well-respected by the administration."

"CMC is a true home away from home. Personally, I think that all the staff here have been nothing but helpful, especially for incoming freshmen, who often have a rough transition time." Students appreciate access to faculty, the small campus size, and the campus living situation, which "promotes good friendships" among students of all classes. "If I had the choice of going to Harvard, Princeton, or Yale, I would still have chosen CMC," says one student. Another, headed eventually to law school, makes the point that "the school helps its students succeed."

What Happens After College

After Claremont, about two-thirds of students go on to graduate or professional education over time, with two-thirds of these to law school and business school, and one-third to other academic programs. Immediately after graduating, however, some 63 percent of students enter the professional employment market. From the class of 1999, 38 percent went right to law school, including some of the top programs in the country (Harvard, Georgetown, Columbia, Berkeley, and so on). Consulting and investment banking are the most popular early career areas, but the list of choices is eclectic and includes nonprofit options like the Peace Corps and Outward Bound and academic fields from education to social psychology. Nevertheless, some 70 percent of CMC graduates have "economics, government (political science), or both in their diplomas." An administrator notes: "We focus strongly on the development of leadership skills—a fact that is reflected in our students' career outcomes: One in eight CMC alumni now holds an executive management or key leadership position in business, industry, the professions, the sciences, government, or public service. CMC alumni achieve the success that they do because they take from campus the ability, skills, and confidence to succeed, as well as the ability to apply information—whether mathematical, technical, scientific, philosophical, or moral—to problems they encounter in their work as well as in their daily lives." Going further, we hear that "CMC graduates have pride in themselves and confidence in their ability to 'take on the world' in many arenas: business, medicine, government, law, and much more. Our students are leaders, not only in the organizations in which they are building their careers, but also in city, state, and national government, and in their local communities."

COLBY COLLEGE

4000 Mayflower Hill
Waterville, ME 04901-8848

(800) 723-3032/(207) 872-3168

www.colby.edu

Number of undergraduates: 1,800
Total number of students: 1,800
Ratio of male to female: 47/53
Comprehensive fee for 1999–2000: $31,580
(includes room and board, tuition, and fees)
% of students receiving financial aid: 40
% of students graduating within five years: 85
Endowment: $243,000,000
Endowment per student: $135,000

Overall Features

Here is another example of a small, independent college founded in the early 1800s to educate its students in the classic liberal arts tradition and encourage them to assume positions of leadership in the larger world. The Northern Baptists who founded the college made sure to include a clause in its charter that promised religious freedom to all who attended. This attitude was expressed later in its admission of African Americans and women more than a century ago. Colby had the first campus-based antislavery society in the nation (1833), was the first previously all-male college in New England to admit women (1871), pioneered the January Plan (1962) (see "Curriculum, Academic Life, and Unique Programs of Study"), and encouraged outdoor orientation trips for all new students. The college's recently

retired president made it an active policy to attract students of differing racial, religious, and ethnic backgrounds. Still a small, entirely residential college, Colby places a premium on its faculty teaching and advising students in a personal and friendly environment. There is also a heavy emphasis on community interaction and sharing. This philosophy led to the elimination of the fraternity system on campus in 1984 and, in its place, the development of a residential Commons Plan. The dormitories were configured into four commons, each with its own dining hall, social activity center, and student-led activities and regulation. Faculty are affiliated with the commons and some even live in a dormitory. Colby students take advantage of the isolated, wintry setting through skiing, club outing activities, and cheering for their Division III intercollegiate teams. A majority of students play on a varsity or intramural team. A good number are attracted to Colby for its strong environmental science studies, which also reflect its country setting.

What the College Stands For

Colby sticks closely to "The Colby Plan," which consists of "10 educational precepts that reflect the principal elements of a liberal education and serve as a guide for making reflective course choices, for measuring educational growth, and for planning for education beyond college. Students are urged to pursue these objectives not only in their coursework but also through educational and cultural events, campus organizations and activities, and service to others." These precepts, which the college believes are at the heart of a liberal arts education, are:

• To develop one's capability for critical thinking, to learn to articulate ideas both orally and in writing, to develop a capacity for independent work, and to exercise the imagination through direct, disciplined involvement in the creative process.

• To become knowledgeable about American culture and the current and historical interrelationships among peoples and nations.

• To become acquainted with other cultures by learning a foreign language and by living and studying in another country or by closely examining a culture other than one's own.

• To learn how people different from oneself have contributed to the richness and diversity of society, how prejudice limits such personal

and cultural enrichment, and how each individual can confront intolerance.

• To understand and reflect searchingly upon one's own values and the values of others.

• To become familiar with the art and literature of a wide range of cultures and historical periods.

• To explore in some detail one or more scientific disciplines, including experimental methods, and to examine the interconnections between developments in science and technology and the quality of human life.

• To study the ways in which natural and social phenomena can be portrayed in quantitative terms and to understand the effects and limits of the use of quantitative data in forming policies and making decisions.

• To study one discipline in depth, to gain an understanding of that discipline's methodologies and modes of thought, areas of application, and relationship to other areas of knowledge.

• To explore the relationships between academic work and one's responsibility to contribute to the world beyond the campus.

These precepts are another excellent summary of the goals and tenets of a liberal education, forming the core of Colby's mission. Like other Hidden Ivies, Colby has seen its student body increase in qualifications and recognition. "Colby was a fine regional college; it is, today, a fine college with a national and, indeed, international reputation. It has changed by an ever strengthening faculty which has brought an increasingly stronger and more diverse student body," says one long-serving dean.

Another administrator notes Colby's perception of itself and its bucolic environment: "Colby's classic, small-town, liberal arts campus creates a community of learning and a rich intellectual climate. The resources dedicated to cultural programming are extraordinary and there is always something stimulating going on . . . The location also makes Colby one of the safest environments in which to study, as campus crime statistics prove." But what makes Colby most unique, raves one administrator, are the people: "It's the personal touch, whether it comes from the president and his wife visiting a sick student in the hospital or the physical plant department helping students build a climbing wall in the field house or secretaries serving cookies at 'Midnight Munchies' exam breaks. When a late-April snowstorm hit while the admissions office had a big group of recruits visiting, students loaned enough Bean boots so the prospective students *and* their parents and siblings all had proper footwear for the campus tour; they all enrolled . . . When the downtown arts center ran into financial difficulty, Colby stepped up with some funding and volunteer support of key administrators to bail it out and make it a nonprofit institution. When, at a local restaurant, our Rhodes Scholar told President Cotter he was taking an unsuccessful finalist for the scholarship out to dinner, Cotter surreptitiously intervened and picked up the tab."

Curriculum, Academic Life, and Unique Programs of Study

Colby insists on student exposure to a full range of the arts and science offerings. While it has a relatively traditional set of requirements that must, as a result, be fulfilled it also gives students some highly appealing means for meeting them. "We have resisted specializing and emphasize the broad liberal arts," says one administrator. "Unlike many colleges which eschewed traditional programs [i.e., foreign language requirements] in years past, we have clung to the broad basics while adding many of the programs that are demanded by the times." For example, to complete the language requirement a student can spend a semester in an intensive language course in Mexico or France. A large number of study-abroad programs are available throughout the four years, and on campus, the college has recently built the Oak Institute for the Study of International Human Rights, which brings a human rights practitioner to campus for a semester and includes a lecture and scholarship program. Colby maintains the January Plan, which many other colleges dropped a few years ago for reasons not fully understood. Over the course of the JanPlan a student can pursue a specific topic in considerable depth without the pressures of the regular academic semester. This can take the form of intense study of a subject with a professor, an internship, an artistic project, or study abroad. Colby students tend to favor concentrations in history, government, international relationships, English, and environmental studies. The recent completion of the Olin Science Center will attract more faculty of note and students to science studies. A large art museum has also been constructed recently. Another interesting program is the Integrated Semester Studies program, which teaches three courses in different disciplines as an interdisciplinary cluster. The administration and trustees have been hard at work on behalf of Colby students in raising money to enhance the learning and living opportunities on campus.

One dean expects to see "increased emphasis on student services. There is no college with a better residential life program." The college sees itself as having strong student services, faculty advising, specialist deans for multicultural and international students, a Writers Center, and a tutoring program. Colby is a "hands-on" kind of place, with "a great deal of individual faculty attention in and out of class and with a student service program that emphasizes close attention and follow-up, which our students appreciate." Another administrator notes that "Colby's faculty has always been good. Today it is excellent. Ninety-nine percent of teaching faculty have terminal degrees in their fields and scholarship is high. A common thread is that teaching comes first, and Colby distinguishes itself with the close interaction between faculty mentors and students. Those relationships are part of a long-standing culture at the college, but they also are supported by programs including a books-in-the-residence-halls program and a take-a-professor-to-lunch program. Senior exit interviews continue to rate student-faculty interaction and collaboration as one of Colby's strengths."

Colby has consistently been on *Yahoo! Internet Life*'s "most wired" colleges list.

"Technology: The college's location in central Maine has changed from a mixed blessing to a strength, in part thanks to technology. Fax, e-mail, the World Wide Web, the proliferation of media (including cable channels), satellite television links (up and down), and videoconferencing all keep students and professors intimately linked with the larger world. We have legitimate national experts on various aspects of the federal government, for example, who are wired into Washington despite living six hundred miles away. We have the founder and editor of the *Journal of Contemporary China* who watches the morning news from China in Waterville, Maine, and keeps his finger on the pulse of various Asian countries. We still have the benefits of Maine's marvelous environment but we now have many more opportunities to interact with the larger world." An alum and current administrator notes that "Colby's campus is often cited as one of the nation's most beautiful and, though the enrollment has grown only slightly, twenty new buildings (more than one-third of the buildings on campus) have been built in the twenty-five years since I graduated. Science students who go on to graduate programs marvel at the access to sophisticated lab equipment (electron microscopes and MRI, for example) that undergraduates here enjoy when, as grad students, they have to wait in line for equipment at major universities. The library has improved from a somewhat neglected facility and collection twenty-five years ago to a larger, better stocked facility that is a technology leader among its peers. The increase in Colby's resources [i.e., endowment] has ensured top-notch facilities that are well maintained."

Major Admissions Criteria

Colby, like its peer colleges, looks at each individual application carefully to find out as much as it can about what makes its students tick. Unlike the other two Maine colleges in the Hidden Ivies, Colby does require the SAT, which it views in the context of an applicant's overall profile. An admissions official describes the criteria that Colby prioritizes in admitting its students: "Colby College looks closely, in a personalized way, at each application. Factors that weigh most heavily in our decisions include a student's high school record, though not only the record of achievement one has accumulated, but, just as important, the degree to which an applicant has challenged himself or herself in the courses he or she has elected to take. The majority of students admitted to Colby place near the top of their graduating classes, after having taken advantage of their high school's most demanding courses. While we have found that a student's academic performance in a rigorous college preparatory curriculum is the single best indication of his or her promise for success at Colby, standardized test scores also may indicate academic potential and will enable us to measure along a common yardstick students applying from all over the world. Colby requires the SAT I. But SAT II scores are optional. Writing that is required on the application helps us to understand each applicant's ability to think critically and communicate effectively through a medium that will be of the utmost importance throughout each student's college career. Required recommendations from college counselors and teachers (2) help us to gain a more comprehensive perspective on each applicant and often influence our judgment. In making admissions decisions, Colby seeks excellence—in academics, art, music, theater, work experience, publications, leadership, public service, or athletics. We value racial, cultural, and socioeconomic diversity throughout the college and seek candidates from all parts of the country and the world."

The Ideal Student

Prospective students who visit the campus seem to know almost immediately if this is the right place for them to spend four years. With its handsome set of buildings on a pristine campus, Colby appeals to a mainstream, socially outgoing individual who seeks an intimate community and culture. A dean describes Colby students as "nice and smart." Those with an interest in an active learning format and the out-of-doors will thrive in this setting. Those looking for an urban experience with lots of off-campus options and a very diverse student population are not likely to choose Colby. "There's still a sense that Colby has a way to go before its student body reflects American demographics, but in the last twenty years minority enrollment has increased from two percent to about ten percent and next year's freshman class will be the most diverse ever. In addition there is another five percent cohort of international students. It remains a priority to recruit more students and faculty of color. A multicultural center was opened three years ago and Colby is aggressively pursuing initiatives to make the campus welcoming and comfortable for all students." Most students will spend a semester or year in one of the many off-campus programs and return to Colby refreshed and eager to reunite with friends and certain teachers. This is not a community made up of tree-huggers, to use the vernacular of college students today. It is more a moderately traditional, socially and physically active group of collegians who thrive in a close-knit community and appreciate good teaching. It is not the best place for loners, urbanites, and social reformers.

Student Perspectives on Their Experience

Students choose Colby because of its academic reputation, location, and relaxed, active environment. "I chose Colby because of the great reputation it had as a work-hard, play-hard kind of school. The close connection with professors and the friendly atmosphere also played a major role in my decision. Colby's campus was one that I fell in love with when I came up to visit as a pre-frosh. It is absolutely beautiful." A sophomore notes, "Unsure of what major I wanted to pursue in college, I was attracted to a liberal arts school. I also knew that I wanted a small school. Visiting the campus is what really persuaded me—I had a great time; all the students were so friendly and the campus is beautiful. Each time I visited I just had a good feeling about the school." Another graduate says, "Not only did Colby fit all of my requirements, but I fell in love with its white columns, brick buildings, ivy, and spacious campus." What students find is a quality academic program, great faculty, and a high-achieving peer group. "I think camaraderie among students pushes others to succeed. The

academic environment is competitive, and the competition helps students to succeed."

Says one 1999 graduate, "Colby does a lot to attract some fantastic professors who really have an impact on one's life. All professors and members of the administration have an open-door policy which really helps a student feel like they belong here. The professors do their own outside research and there is ample opportunity to apply for a research-assistant position. I never would have thought that I would be working for a professor after my first year of college, but it happened." And, "The close attention from the professors really helps a student succeed. If you are struggling, the professor's door is open to you, along with possible tutoring from a TA. Furthermore, the open-mindedness of most of the students is a big plus." Another student says, "Colby offers academic help to students in many ways. I can offer personal examples. Struggling in our biology class, my roommate and I found it easy to meet with a tutor who was very helpful and accessible. My calculus class met with two math majors once a week to clear up any questions dealing with our homework. I frequently met with a TA for guidance in economics projects and my English comp. professor scheduled regular one-on-one meetings to discuss individual writing issues. Help is readily available to students and they are encouraged to take advantage of this help."

Students love the JanPlan options and the COOT program: "This is Colby Outdoor Orientation Trip. Each incoming student is put into a group of about ten other first-years and two or three upperclassmen. These groups hike, canoe, bike, and/or camp out in the beautiful Maine woods for four days before on-campus orientation and classes begin. COOT provides new Colby students with a core group they come to depend on in the first few weeks of school, which is usually a very critical time. Most students remain in contact with their COOT friends throughout all their years at Colby."

Students recognize the need for more diversity on campus, and, in line with their administrators, the importance of continued endowment building to support the college's programs: "One of the toughest areas for Colby is their ability to raise money. There has been a recent push, but Colby is still lagging behind other schools in this category. In addition to fund-raising, Colby has tried to bring in more minority students. Although the numbers are increasing, it is a slow process that will take many more years to get to the level they want to be at." And, "Colby remains very homogenous with its predominately straight, white, middle- to upper-class population—despite efforts to diversify. Offshoots of this problem include poor town-gown relationships, self-image problems among students (eating disorders and alcohol abuse are two symptoms of these problems), and a generally elitist and apathetic student body." Nevertheless, students report an accepting environment for racial/ethnic minorities, international students, and gay and lesbian students. And "there are plenty of cultural activities on campus—theatrical performances, symphony and choir concerts, festivals celebrating various cultures, and lectures."

Who does well at Colby? "The students who work hard and play hard do the best," says one student. Another reflects, "It is difficult to say who does best because there are many ways to succeed at Colby. My grades are pretty good, but I am not a straight-A student, and I'm a pretty good runner but not a huge point-scorer for the track team. If grades or athletics are what is important then you could say that I am doing just pretty well. But I did declare a major, accomplished many goals along the way, and made friends and memories that will last a lifetime. Everyone does best at what they feel is important." A graduate adds, "Colby provides incredible opportunities. I was able to play three sports, work for the school newspaper, write Web pages, and obtain interesting internships. I also was able to study abroad. So Colby offers opportunities that couldn't be found at larger institutions."

"There are some exceptional and very special people who work on Colby's staff [custodians and secretaries]; the food and dining hall are far superior to any I've experienced elsewhere; some members of the faculty and administration are exceptional people and mentors; and the library is phenomenal," says a recent grad.

What Happens After College

Insists one dean, "Eighty-five percent will go to grad school. They are prepared to do and be anything they want to do or be. I can prove it." Another administrator notes, "Colby's commitment to the liberal arts embraces the firm belief that the breadth and quality of a Colby education should be extended to include an equally broad choice of meaningful and rewarding career opportunities. The Office of Career Services strives to acquaint students with career options, offers insight into various professions, and assists in preparation for the actual career search." About 15 percent of Colby grads begin graduate school within three months, and within three years some 40 percent have entered or completed graduate school. Choices include study in a broad variety of arts and sciences programs, law school, medical and veterinary school, and business school. Employment directions are also broad. Students go on to work in education (22 percent); economics, sales, and consulting positions (20 percent); science and health (16 percent); financial services (12 percent); social services and government (12 percent); communications (10 percent); and law (8 percent). "The Peace Corps and AmeriCorps are also popular destinations," notes an administrator, "perhaps in part stemming from an active student-run Volunteer Center." She goes on to say that "the assistance of alumni and friends of the college is particularly important. With the generous support of Colby graduates and parents of current students, a broad network of persons in various professions and widespread geographical locations has been established to assist students and alumni in career exploration. Parents and alumni have agreed to conduct informational interviews, be hosts for on-site visits, sponsor internships for January and the summer, and provide housing for interns and job seekers in their areas."

COLGATE UNIVERSITY

13 Oak Drive
Hamilton, NY 13346-1383

(315) 228-7401

www.colgate.edu

[
Number of undergraduates: 2,760
Total number of students: 2,760
Ratio of male to female: 49/51
Tuition and fees for 1999–2000: $24,750
% of students receiving financial aid: 44
% of students graduating within five years: 87
Endowment: $344,000,000
Endowment per student: $124,637
]

Overall Features

Colgate may call itself a university, but it is in reality an undergraduate, residential liberal arts college. Here is another of the outstanding colleges established by the Northern Baptist Church, this one in 1819 to train young men for the ministry and civic leadership. Colgate reflects the more traditional model of the private, residential college that prevailed prior to the social revolution of the late sixties. It admitted women in 1970, as did many other all-male colleges at that time. While it has expanded its curricular offerings extensively over the years, Colgate still requires students to undertake a core curriculum and fulfill a number of subject distributions in order to graduate. The strong commitment to outstanding teaching is a hallmark of the school. Here is a refreshing example of a faculty whose members are evaluated regularly on the quality of their teaching as well as on original research in their respective disciplines. Teachers are expected to engage with students out of the classroom as well in this intimate, rural campus setting. The workload is pretty intense and students are expected to participate actively in the classroom. The Greek tradition has been a signature feature of campus life for many years. Close to one-half of the men and a third of the women belong to fraternities and sororities respectively. Many of them live in their chapter houses instead of dormitories. Not unlike many other institutions, Colgate finds itself in the throes of heated debate on the role of Greeks on campus life. The faculty recently voiced its opinion that they should be removed since they seem to be a socially divisive element and the main source of anti-intellectualism on campus. Both undergraduates and alumni are highly emotional on the value of the system and will not let the trustees and administration eliminate them without a fight. A final resolution on the future of the Greek system is still to be determined.

Athletic competition at the Division I level is a significant component of the Colgate experience. Talented high school athletes who want to combine an excellent education of a traditional nature with strong athletic competition find Colgate particularly attractive. Its location in rural, upstate New York has created a self-contained community of smart, physically active, socially engaged students who can thrive on its beautiful campus. It has also resulted in a mostly middle- and upper-middle-class cohort of students who feel comfortable in this traditional environment. It is a challenge for Colgate to attract a critical mass of minority students to the campus. The great majority of students prepare for professional careers in business, law, and medicine, befitting their traditional backgrounds.

What the College Stands For

A dean describes Colgate's educational goals: "We are dedicated to the total education of students, both academically and personally. We are proud that students learn in the classrooms, labs, studios, study groups, and other traditional academic settings. But at Colgate we recognize that significant learning also takes place in the residence halls, on the athletic fields, in student activities, and other nonacademic settings. We work hard to create vibrant learning environments in all of the college's programs, and we think we do a pretty good job." Another dean notes that

"Colgate is committed to providing a rigorous and far-reaching liberal arts experience for undergraduate students. A Colgate education demands that students stretch their limits and engage with ideas in a variety of different settings."

Residential life at Colgate has been improved, even to the point of the college advertising for volunteers to move off campus to relieve a temporary housing crunch last year. The college has added and enhanced facilities, moved fraternity and sorority rush to fall of sophomore year (students may not live in Greek houses until junior year), and increased to 15 percent the proportion of students of color on campus through recruitment and retention efforts. "Improving our village's economy and services to residents and students is a high priority for Colgate," says one dean, noting the college's goal to help alleviate the results of a sagging economy in upstate New York. What makes Colgate unique is "our commitment to the overall growth and education of students, study groups run by faculty, a world-class career services center, division 1-AA sports, our location in upstate New York, our willingness to experiment with change, and the amount of technology available to students." With 2,760 students, Colgate is larger than most other small liberal arts colleges, yet still retains the "feel and personal touch" of its smaller cousins. The college sees its students as having become more academically serious and qualified over the past decade and less inclined to drink. "We still have a ways to go in terms of reducing the role of alcohol use in student life, but it is far less central than it was ten years ago. Accordingly, there is a more vibrant intellectual life outside the classroom than there was at that time, which makes Colgate more attractive and comfortable for students with social interests that vary from what has long been the 'norm' here."

Several administrators stressed their belief in both the college's excellence and its refusal to ever "be satisfied with the status quo." A dean illustrates Colgate's view of itself, as well as one of the central arguments of the Hidden Ivies: "I have worked at two schools ranked higher than Colgate University by a news magazine, and my children attended two others. Having worked at two other places and being a consumer at two more, I seriously question numerical rankings. Overall, I believe Colgate University is superior to three of the other colleges and on par with the fourth. We care deeply about our students and our programs reflect that attitude. I feel honored and humbled to be the dean of such a fine college."

Curriculum, Academic Life, and Unique Programs of Study

Colgate does not apologize for retaining its traditional, Western civilization–oriented curriculum and requirements. The administration instituted in recent years a set of core requirements for first- and second-year students that include these four major courses: Challenge of Modernity; Western Traditions with Literature; Cultures of Asia and the Americas; Scientific Perspectives of the World. In addition, two courses each in the humanities, sciences, and social sciences are required. This is an effort to make certain all graduates will have been exposed to the major disciplines of a liberal arts program. The most popular majors once these foundation requirements have been fulfilled are history, political science, economics, and government. Says one dean, "We expect students to be well-versed in the history and traditions of Western thought and require them to engage in the study of at least one non-Western culture. Our core curriculum also requires students to explore the role of science in contemporary society, and the distribution requirement ensures that students may not be too narrow in the way in which they think about and engage in the academic endeavor."

The college has recently enacted an honor code that echoes those of many other Hidden Ivies. Other key programs are study groups and student research. Numerous support programs and organizations are available through the ALANA Cultural Center. "Theme houses" like the Harlem Renaissance Center, Bunche Peach Studies House, and Ecology House appeal to students with particular interests and affinities. "Clearly," says one dean, "supporting students who are different from the typical Colgate-bound student is a high priority for us." The college runs the OUS program and Skin Deep retreats to help students become acclimated to campus. All faculty at Colgate, a point of pride for the college, are involved in student advising, as are a network of deans and administrative advisors. Again emphasizing Colgate's interest in having its students succeed outside as well as inside the classroom, some 75 percent of students are involved in athletics of some sort, and the college supports over 100 student activities.

Major Admissions Criteria

Colgate's admissions officers are clear about their most important criteria in admissions:

1. As with most of our peer institutions, the #1 criteria for admission is a strong high school transcript. If we had only one document to use in our selection process, it would be the transcript. It records a student's academic achievement, ambition (seen in course selection), and pattern of success, advancement and diversity of interests over 3–4 years of high school. From our experience, a student's high school records (GPA) and the level of difficulty in their course selection combine as the most powerful predictor of academic success in the first year at Colgate.

2a. Next in importance are recommendations and testing. These items "tie'" for second place because they really do counterbalance each other as elements that broaden our understanding of a student's achievement—in the eyes of those who know them at school and according to a standardized measure of achievement. . . . We're looking for the stories about a student in the context of their high school, their classes and the school community. We want to know what contributions the student makes, in and out of class, and whether their ideas, involvement and example have an impact on their peers . . .

2b. Standardized testing is also important. For most selective colleges, standardized test results will play a significant role in the admission process. That does not mean that the SAT is *the* measure of who gets into college, just as our saying "the transcript is more important" does not mean that SATs or ACTs can be discounted . . . Colgate takes the time to review every item in an applicant's folder and make an informed, careful judgment about the student's academic preparation and achievement, individual circumstance and accomplishment, and their relative position in our applicant pool. . . .

3. The third priority would be a student's activities and overall "personal profile" of honors, involvements and summer pursuits. These elements are still significant in our decisions, especially at a place like Colgate where student participation in campus life is pivotal to the vitality of the community. We attract (of necessity) students who are "do-ers," who are accustomed to getting involved, making things happen and helping to create the life and activity around them. Ours is not a passive campus, though I would also note that students need not be "super achievers" in order to be admitted . . . We are looking for the individual "story" that is told by a student's choice of activity, and the pursuits to which they dedicate themselves over time—demonstrating a commitment to a particular interest and hopefully a level of achievement or recognition in

areas most important to them. Like the pattern of course selection, this pattern of involvement helps us make a projection about the success and involvement a student might have at Colgate."

The Ideal Student

Those high school students who are seeking an academically rigorous, more traditional orientation to the study of the arts and sciences where good teaching is recognized and rewarded will find Colgate very appealing. They must also feel at home on a campus that favors athletic engagement and lots of social interaction through Greek and dormitory life, as there are few alternatives to be found off-campus in Hamilton and its environs. Colgate students describe themselves as intelligent, confident, energetic, moderate to conservative, career-oriented, and in search of a good time to balance the hours spent on school work.

It is hard to do better in matching the college to its ideal students than the following statement from a senior dean: "We seek students who are demonstrated academic achievers and involved community citizens . . . who are excited by an atmosphere that emphasizes participation—in and out of class—and expect students to engage with one another and their professors. We seek bright, intelligent, engaged learners. I sometimes use the phrase 'practical intellectuals' to suggest the joining together of intellectual life with 'real-life' settings for learning . . . Colgate appeals to the best and brightest students who also seek a down-to-earth and rather lively environment . . . I don't mean purely social involvements, but rather a lively sense of involvement and direct contribution to the college community. Every student at Colgate can make a difference. In a small-town setting, every voice matters. We appeal to students who see the merit in that atmosphere versus a more anonymous, typically larger or more urban setting. Colgate students often speak of the balance in their lives—not a 'perfect' fifty-fifty proportion of academic and personal pursuits, but rather a sense that both sides of that equation really do matter and are worthy of attention. Colgate students and Colgate as an institution are nothing if not 'ambitious.' We sometimes use the word 'scrappy' to describe the institution—suggesting a sense of 'not settling for second-best,' or the 'we-try-harder ethic,' or in our athletics program, the image of a 'giant-killer'—the small school that competes against—and beats—institutions more than twice its size. There is a strong spirit and personality to Colgate, a vibrant, positive, upbeat style. That 'can-do' attitude (corny as it sounds) permeates much of what we do, and characterizes the kind of student who will do well here."

Another dean adds, "We are looking for academically motivated students who have a proven ability to succeed in a rigorous environment, students who are intrigued by ideas and display genuine intellectual curiosity. Additionally, given our small, rural setting, we want students who are motivated to create opportunities for themselves and their peers outside the classroom. We provide resources and guidance, but we want students who will seize the opportunity to edit the student newspaper, or start a programming board, or engage in community service, or direct a play.

The students who do best here are those who extend their zeal for learning beyond the classroom and make meaningful commitments to co-curricular involvement."

Student Perspectives on Their Experience

Students tend to feel "right" on Colgate's campus fairly quickly. "When I stepped onto Colgate's campus for the first time during college visits, I just had this remarkable feeling, a feeling that I hadn't felt at any other school. From that point on it was an easy decision, and I chose to apply early decision. Colgate is unique in that it excels in so many areas. Whether it is academics, athletics, student lifestyles, or even being ranked the number-nine party school in the country, whatever it is, Colgate is consistently found near the top." Another student says, "Upon visiting, I simply fell in love with Colgate. It was the feeling emitted by the students, the small, close-knit community, and the intense academic experience I sampled in the class I sat in on. Three years later, it's still kind of hard to put into words what Colgate means to me, but the maroon runs through my blood. I love the sense of tradition and the fact that Colgate never leaves one even after he graduates here. We produce strong, intellectual people and we develop leaders. Graduates are social beings who are ready to go anywhere in the world and into any field and excel. It's an exciting place to be, the place I have called home for three years, and I can't wait to be a proud Colgate alum after I graduate from here."

A recent graduate "encourages all those who are interested in the university to take a visit and stop a person at random on their way to one of Colgate's beautiful buildings and ask them why they are at Colgate. I guarantee you the answer will be different for each person, but no matter what, they will feel that common bond to their alma mater and so can you. Colgate is the place for those who want to learn, think, and give back to society."

Who succeeds at Colgate? One student says: "Academically, I feel that anyone who takes time out to care about the classes and the intellectual environment here does well. You have to work hard to do well (B+ or A) and that means studying a lot and doing reading and research. If you blow off your classes and don't take your work seriously, you will find yourself in a deep hole before the end of freshman year. Yet, even if you don't get all A's, that's not the point. I believe the opportunity to learn so much from a gifted faculty and a strong liberal arts curriculum lets one explore new areas. It aids one's intellectual growth and that is something no one forgets. Socially, people from families with college backgrounds will have an easier adjustment because they know what to expect. White students of upper-middle-class backgrounds are the norm here. But students who are ready to get involved and be engaged, people who are eager to be social and work with others will excel. Finally, it is a great asset if one can expand his or her mind and comfort area to accept new people and new ideas." A 1999 grad says, "Student leaders who are highly motivated, intelligent, and willing to contribute to the community tend to thrive at Colgate . . . Students who are willing to learn from other people's differences also do quite well."

Students feel supported and valued on campus: "There is a huge support service here. From students and peers involved in Residential Life and the Link staff to the academic and administrative deans, there are a ton of people to seek help from. Every single activity and group is easy to get involved in; it all depends on what interests an individual. I jumped right into student government and student activities by my second semester and was the head of a campus organization as a sophomore . . . Being social and ready to meet and work with people is a huge asset. Also, being responsible and developing good time-management skills is important to balancing a social and academic life. Also, not being afraid to talk to professors and people about problems or questions will help one understand that there are many students who are uncertain about life in and after college. It takes time and strength of personality as well as a positive outlook and some vision added in for good measure because there is so much to offer here for everyone, no matter what you like and who you are." Another student notes that "the Office of Residential Life was an integral part of my experience . . . Res Life offered me the chance to excel as a peer counselor and advisor, through resident advisor and head resident roles. The club sport program allowed me to form my own martial arts club, which taught karate to students, faculty, administrators, and local residents."

Another student says, "I love Colgate for what it symbolizes. The deep sense of tradition that comes during Freshman Orientation when the students walk up the Hill for Convocation in the Chapel—and again as seniors when we march back down the Hill and surround Taylor Lake to symbolize the complete circle of Colgate. The look and feel of the buildings. I love the pride that students feel toward this school. You graduate Colgate with a part of the school embedded in you forever. I am proud of the great commitment to academics, the strong liberal arts curriculum, and the rigorous academic experience that molds us into intelligent young adults. The great tradition of athletics here and how well we perform at the Division I level. The active student body where it seems that everyone is involved in something, from student government to outdoor education to the newspaper or the Greek system. And the sense of community that flows from the people here."

While Hamilton—and the college—could be bigger for some students, "the many opportunities for involvement both within the Colgate community and in the village of Hamilton are invaluable experiences that have taught me just as much as I could learn by obtaining any degree here." As with other similarly located campuses, diversity is an issue at Colgate, but in general the college is respected for its efforts: "Colgate is not by any means an overly diverse campus, but I feel that it does a fair job attracting students of all different races. Colgate has unique housing opportunities available for those students who choose to live in a residence hall that is made up of people of the same race. Also, different student organizations exist that enable students of any race to join who are interested in learning about a specific race. But I'm not sure that Colgate does a tremendous job of providing students of different ethnic backgrounds many opportunities to 'express themselves' and their culture on a regular basis. I feel a lot of that is due to the small rural setting we're found in and not so much ignorance on the part of the university." Another area of improvement could be town-gown relations: "In my opinion, Colgate could do a much better job dispelling the animosity between students and faculty, as well as townspeople, that sometimes surfaces at this small university . . . It really makes a difference if the students feel welcome when they're off-campus. Student-community relations have the ability to add or subtract an important aspect of a well-rounded college education."

"The financial aid programs at Colgate have been huge because I *never* would have been able to come here if it weren't for the great aid package I received, and I know Colgate does its best to provide for all incoming students. I hope I become a millionaire someday so I can continue the tradition of giving back to my school to make sure that kids like me have this chance. The study-abroad program was one of the best experiences in my life. As a kid, I never dreamed I would be able to live in London, have class across the street from Big Ben, see the Eiffel Tower in Paris, and travel to Ireland (a great accomplishment because of my Irish heritage). I was able to immerse myself in the study of British history and culture and I will never forget the people I met or the experiences I had. Study abroad gave me a new perspective on the world, my American identity, and what is really possible for someone who sets his sights high."

"The beauty of the campus is unparalleled," says a senior. "The small Hamilton village and the campus itself are some of the most serene, peaceful places I have ever lived and my experience of spending my summer here only heightens that fact. This school does a good job of promoting community. The residential life is great, and starting in one's first year, the dorm life produces great friendships. Also, the bond between students and faculty is incredible. I have generated strong relationships and even friendships with some of the faculty here, brilliant men and women. The ability to sit down and talk with them, interact and learn from them have been instrumental in strengthening my intellect. Colgate does a great job of creating strong social life. Be it through the Greek system, or, as in my case, through involvement in student-led activities, there are almost too many places for students to explore and get involved in. It is there that I have become a leader, someone capable of meeting people and making things happen on campus." What do students gain through their education at Colgate? "The greatest thing about this place is that I can do whatever I want with my degree. I feel prepared to conquer new challenges and use my liberal arts background to pursue my interests and, ultimately, do a job I love." A 1999 grad with plans to pursue a teaching career (perhaps at Colgate) in psychology wants to maintain academic connections to the college, as "my advisor and I are in the process of publishing my Senior Honors Research Thesis."

What Happens After College

"Colgate students go everywhere after they graduate," says a dean. "Large numbers pursue graduate study or enter corporate training programs, as would be expected. Others travel, work for non-profits, start their own companies, play professional sports, pursue careers in the arts, and the list goes on. The idea that a liberal education prepares people for a wide variety of careers and postgraduate endeavors is not just admission-brochure copy at Colgate. It is our reality." The college notes that, "The Standard & Poor's Executive/College Survey, prepared by the S&P's Register of Corporations, shows that Colgate is among the top two undergraduate liberal arts colleges to supply alumni for senior management positions in the nation's largest corporations." Another dean argues that Colgate students "are prepared for everything! Our students are sought after by all the major corporations . . . The recruiters know our students are smart enough to learn anything in the field and socially 'savvy' enough to become leaders/managers." Alumni connections are also there for graduates: "People become very attached and one of the special attributes is the commitment of our alumni. They truly 'bleed maroon' and are ready to step forward to help those in the Colgate family. Our students have many 'doors opened' in the real world by alumni who are eager to help. Colgate fosters a passionate attachment to the college."

COLORADO COLLEGE

14 East Cache La Poudre
Colorado Springs, CO 80903

(719) 389-6344

www.cc.colorado.edu

Number of undergraduates: 1,955
Total number of students: 1,955
Ratio of male to female: 46/54
Tuition and fees for 1999–2000: $21,822
% of students receiving financial aid: 57
% of students graduating within five years: 84
Endowment: $317,120,000
Endowment per student: $162,210

Overall Features

Founded in 1874 as an independent, traditional liberal arts college on the New England college model, Colorado is an old school by Western standards. Over the past quarter century it has advanced dramatically its goal of enlarging its reputation as a top academic institution with a national student representation. One senior faculty member attributes the many positive developments to "a deliberate search by the college for excellence." It is well on its way to achieving national prominence as measured by the quality of education it provides; the credentials and reputation of the faculty; the multitude of new academic, recreational, library, science, arts, and athletic facilities; and the selectivity of the student body, with its diverse geographic backgrounds. Of the students, 75 percent come from outside the state, including all fifty states and thirty

other countries. Both the Eastern and Western states are well represented. The president states that for 125 years Colorado College has prospered by blending its commitment to tradition with its willingness to embrace innovation; that it has implemented its core values through small classes, personal interaction between students and faculty, and academic rigor.

In 1970, the college made its most momentous curricular change in delivering the liberal arts education it still viewed as the heart of the educational experience with its initiation of the Colorado College Plan, familiarly known as the Block Plan (described in detail later). The only top college in the country with this unique academic structure, Colorado helps to attract strong students to its residential campus snuggled in a wide valley with a commanding view of the Rocky Mountains and Pike's Peak. With its magnificent setting in one of the most popular regions of the country for college-age students and a dedicated faculty leading the Block Plan, the college sees itself as possessing the potential to compete with the most selective of the liberal arts schools we include in this book. A senior administrator describes the essence of the college this way: "We offer a major liberal arts education with an altitude!" Given the underlying philosophy that critical thinking, creative expression, and communications skills are best developed by close association between teacher and student, the Block Plan requires teachers to engage actively with students in small classes and seminars and in the field work that is a frequent component of the course work. The ratio of students to faculty is eleven to one and it is a set rule that no class can exceed twenty-five students. The average class holds fifteen students and the majority are conducted in seminar fashion.

Colorado College's location in this distinctive area of the American West also influences the educational environment, with special majors and field work built into many courses and concentrations. The campus is set in Colorado Springs, the second largest city in Colorado, seventy miles south of Denver, the largest. Although there are some cultural activities available in the conservative yet ethnically diverse Colorado Springs, the majority of students spend their free time on the nearby mountains, trails, and waterways, or motor up to Denver where there are more fun outlets on the weekends. Students are more likely to be seen with their inline skates, skis, climbing ropes, and mountain bikes than with the traditional accouterments of Eastern college students.

The college leaders are committed to expanding the mix of undergraduates to represent the national demographics. Right now the percentage of Hispanic and black students is quite small. Almost every admitted candidate who demonstrates need is offered an attractive financial aid package, and the college offers a few awards to needy international students. The director of athletics believes one of the valuable roles the athletic program plays is in the recruitment of students of color and diverse cultural and economic backgrounds. The program provides the support of a defined social unit as students make the transition to their new community. The athletic teams have a clearly stated policy of inclusiveness and any form of discrimination will not be tolerated. Athletics are a major component of life at Colorado College for a majority of students either through very competitive Division III play or intramural teams. The athletic department believes strongly that its programs make a major contribution to the academic and social development of their students. The men's hockey team and women's soccer team compete very successfully at the Division I NCAA level, and eight men's and women's teams compete in Division III intercollegiate sports. Three-quarters of students participate in intramural sports.

There are many outlets on campus for artistic and literary interests for the less athletically inclined. Students describe themselves as a liberal-thinking group with a relaxed living and dress style that can belie the heavy workload required under the Block Plan. The many special interest groups that represent particular gender, political, and ethnic groups found on virtually all campuses today are present, but on the whole they seem to shy away from the aggressive and vitriolic behavior present at the more liberal colleges. This is in keeping with the individualistic nature of the Colorado College community. Studies are demanding but the Block Plan allows for set breaks for traveling, relaxing, community service, and getting out to the mountains. Students are required to live in campus housing for their first three years and are guaranteed housing for the fourth year. As a result, about a third of students live off campus at any time. While there are recognized fraternities and sororities, a very small proportion of students choose to join them. Special interests and theme houses, such as language and arts specialties, are also sponsored by the college.

What the College Stands For

The president of the college supports the intellectual and personal goals as expressed in the mission statement: "To provide an excellent liberal arts and sciences education which challenges well-motivated students of varied social, ethnic, and economic backgrounds and prepares them for positions of professional leadership and civic responsibility in an interdependent world." Underlying this mission are four priorities: enriching student intellect through high expectations for imaginative teaching, learning, and scholarship; promoting understanding of both similarities and differences among people with diverse backgrounds and interests; collegiality and mutual respect; and involvement in and responsibility to one's community. The administration seeks to provide a transformational experience for its students that they

will carry with them for their adult lives. The college believes that only by a continuous process of looking at itself critically and creatively will Colorado continue to innovate and meet its goals in providing a first-rate educational and developmental experience. Said a senior administrator: "The college has sought the very best among students, faculty, and staff, and has looked in particular for individuals who believe in the values to which the college is committed. The success of the college is a testament to the extraordinary individuals who have come here. Our faculty is focused on instilling curiosity, wonder, and intellectual excitement in the young women and men who come to study here. This focus, along with small classes and the intensity of the Block Plan, helps students succeed in their academic pursuits and gives them the confidence to explore and discover." The administration is committed to broadening its economic and cultural mix of students and faculty by active recruitment and retention of minorities in order to enhance the environment of the campus and the growth of all students. The willingness of the college to embrace change over time if it betters the way students learn is articulated in these words of the president in her 125th anniversary speech: "I envision several enhancements that build upon our core values: every student would complete an original creative project, from collaborative research with faculty to documentary filmmaking; every student would have an intercultural experience, whether in Kansas City or Kyoto." A professor's view from the classroom offers additional insight into Colorado's mission and style: "In my view, the only advantage—but it is a huge advantage—that a small liberal arts college has over other types of colleges and universities is our personal approach to education. Small liberal arts colleges like Colorado have the ability to engage and challenge our students as no other type of college or university can. For me, the most tangible sign that I'm accomplishing my goals is when a student of mine has the 'a-ha' experience and I can visibly see that the student has recognized some new truth or come to some new realization—a truly gratifying moment. I've seen this in the classroom many times; it recently happened in a class as we were discussing the potential of the Internet to transform traditional business practices. I've also seen it happen recently to a student who was talking with a visiting alumnus of the college. During this conversation, the student arrived at a new appreciation of the vast array of career opportunities awaiting him, and he came away totally transformed."

Curriculum, Academic Life, and Unique Programs of Study

The Block Plan is the most distinctive academic feature, and it differentiates Colorado from all of the top liberal arts colleges we feature here. It basically works this way: Students undertake only one course at a time for 3 1/2 weeks with a professor who teaches only that course. Some courses may last for one block while others for two or three, depending on the content of the subject and amount of material to be covered. Students take eight courses or blocks each academic year. This allows both students and faculty to focus all of their attention on the one course. Another

unusual element of the curriculum is the way students bid for their desired courses on a point basis. The more points they bid for a favorite class, the more likely they are to win a place, since all classes are kept to a maximum enrollment. Many courses will combine a heavy reading and writing load with field work, since there are no other conflicts in the schedule.

The college describes the flexibility of the Block Plan as one of its major advantages. "Each class is assigned a room, reserved exclusively for its faculty and students, who are free to set their own meeting times and to use the room for informal study or discussions after class. Since competing obligations are few, time can be structured in whatever way is best suited to the material. No bells ring. Nothing arbitrarily intrudes after fifty minutes to cut off discussion. An archaeology class can be held at the site of a dig for one block, followed by the second block for laboratory analysis. A biology class might have a week of classroom orientation, then go to the field for two weeks. An English class can spend one morning reading a Shakespeare play aloud and the next morning discussing it or getting together with an acting class to try a few scenes." The college runs a second campus in the San Luis Valley, 175 miles from the main campus, where students can reside as part of their focus on Southwest studies or other specialized topics. You get the idea. It is common for classes to extend the classroom walls to the Grand Canyon or to Indian reservations or even abroad. With only one subject to cope with, students are free to create independent study or research projects or self-development study such as learning a language, a musical instrument, community service volunteering, or working on their athletic and out-of-door interests. There is little questioning of the intensity of the academic work since the content of a traditional full semester is covered in three and a half weeks. Students cannot hide from their teacher and their peers. The assigned work must be completed and contributing to the class discussion is a given. At the end of each block there is a four-and-a-half-day break for students to do as they wish.

The faculty are known to work as hard as their students to make such an intensive learning structure achievable. Colorado College boasts a faculty of 160 full-time professors, forty-eight part-time professors, and another fifty who visit each year to teach a course in their specialty. The Block Plan also makes it feasible to offer a wide range of interdisciplinary concentrations. These include, but are not limited to, environmental studies, international affairs, North American studies, Southwest studies, American ethnic studies, and studies in war and peace. Colorado students must fulfill general education requirements that are rather flexible: two courses in the Western tradition, three in the natural sciences with a lab, and two additional courses in non-Western or multicultural topics. First-year students are required to elect a theme-based seminar that combines coursework with outside activities to broaden their exposure and hone their reading and writing skills. One-half of the student body typically takes advantage of off-campus terms for one or more blocks. Students can be found in New York, Paris, Florence, or in any major mountain range doing field work in geology or environmental matters. Approximately 55 percent of all students par-

ticipate in study-abroad opportunities in every part of the globe on Colorado's own sponsored programs, and a multitude of off-campus programs is sponsored by the Associated Colleges of the Midwest, a group of quality independent liberal arts colleges. For those already focused on their future plans, there are pre-professional programs in medicine, dentistry, law, physical therapy, education, and veterinary medicine. Reflecting its environs and some of its student body, a major in health fitness and exercise is offered. Cooperative 3/2 degree programs in engineering are offered with Columbia University, Rensselaer Polytechnic Institute, the University of Southern California, and Washington University, and a 3/3 degree program with Columbia.

Major Admissions Criteria

Because the Block Plan is central to the learning experience, the admissions committee puts special emphasis on signs of a candidate's commitment to engaging in a challenging academic environment, one that places great weight on writing and communication skills, on potential for contributing to the small classroom teaching method, a genuine involvement in the liberal arts subjects studied in depth in the Block Plan, and the ability to complete a demanding workload in a timely fashion. The faculty express their major interest in students "who are truly interested in something, students who are willing to work hard at something not because someone else thinks they should, but because it is something they really want to do well." These abilities are best measured by the content and level of the secondary school curriculum.

Special attention is paid to students who complete advanced placement, honors, accelerated, or enriched courses. Class rank is one indicator of performance. Over 80 percent of admitted students rank in the top 25 percent of their graduating classes. The application essay and samples of writing ability are given weight, and teacher recommendations and test scores are also taken into consideration. Special talents, extracurricular participation, and international experiences are important predictors of major contributions to the small and intimate community that characterizes Colorado College. Geographic, multicultural, and intellectual special interests are strongly factored into the decision process. Colorado has a substantial scholarship budget that is utilized to recruit outstanding students of all backgrounds. Thirty athletic scholarships are given to top-flight men's ice hockey and women's soccer players. Special merit scholarships are available for students who plan to major in the natural sciences.

The Ideal Student

An atmosphere of individualism within a supportive community and a kind of Western, open-plains philosophy informs the environment at Colorado. A longtime professor of English describes what she believes are the qualities of the successful Colorado College student: "A good academic background certainly doesn't hurt, but I believe attitude may be more important. I particularly like students who ask questions, explore new areas, welcome new ideas, cooperate with other students in class, and work hard in extracurricular activities." The direction

toward grade-grubbing and competition among students rather than an enjoyment of creative and critical learning is not something she hopes to see develop as the college has become increasingly more selective. Another administrator who works closely with students agrees with what kind of students do well at Colorado College: "Intelligent, self-assured, broad-thinking, extroverted, and possessed of plenty of work ethic and initiative, and a commitment to excellence." It should be clear to a potential candidate that four years at Colorado College will provide a fresh and dynamic experience. It should also be evident that to succeed in the Block Plan approach to learning, one needs to have good study and time-management skills, strong reading and writing skills, and a desire to immerse oneself in a subject and work closely with the faculty and fellow students. The academic program is intense due to the short period of time available to master a subject. Students say that it is especially demanding for science majors to stay on top of their work. The opportunities to combine theory and hands-on learning cannot be beat for those with real intellectual curiosity and the ability to focus. While there is a generally liberal flavor to the political life on campus, there is little in-your-face behavior. There is a mix of conservative and liberally inclined students to make the community work for just about anybody. A "sense of adventure" and a love of the out-of-doors and more physical space to stretch one's legs and mind make Colorado College a terrific place for an active and involved student. Any strong high school student considering the top Eastern liberal arts colleges would do themselves a favor by looking into the college that officials refer to as "the only national liberal arts college in our time zone."

Student Perspectives on Their Experience

A sophomore speaks for most of her peers: "I chose CC because of its special and convenient schedule (the Block Plan), its beautiful location, its reputation, and the size of classes, which allows for more personalized attention and better chances for individual success." Those who do best are "the people who are committed to their success and do not procrastinate. The ones who take a wide range of courses, and pick their classes according to professors and not according to sexy titles. The people who put all their efforts into learning what the professor is teaching, and trying to get the most out of each class." A senior points to "students who are willing to go, go, go and then totally relax on block breaks. It's an all-or-nothing system—you either work all the time, or, during block breaks, not at all. Students who learn to write well, who don't fall behind in their classwork, who are willing to work hard on a day-to-day basis during a block, do well here." Students like the Block Plan overall and selected CC because of it, but they emphasize the need for potential students to be very aware of the plan's demands and proactive in determining their path through the college. Says a senior, "The Block Plan is a wonderful way to conduct college classes. Especially with a limited class size, students are able to dive into a subject completely—there are stories of students who begin to dream about their class topic." A

junior adds: "The Block Plan really allows me to get to know my professors personally. The classes must be small and we see each other every day. I love it."

Students say that Colorado is diverse, but only moderately so, and many would like to see the college continue to expand its efforts to attract, retain, and aid multicultural and international students. A junior resident advisor says, "All kinds of diversity exist on our campus, it is just not always visible to the naked eye. The college always seems to be looking for ways to make students with different backgrounds feel welcome, but people always complain. As a head resident I am required by the college to lead at least two activities that focus on diversity issues each semester. This has gotten some healthy conversation going in our house, but the campus as a whole is full of complaints about not enough diversity and not enough programs for students here who are from diverse backgrounds." Some would like more of a voice in governing their affairs and working with college officials on student life, while others wish the college did not have such a "laid-back" attitude. Students appreciate their opportunities for study abroad and community service close to home. They generally feel supported and connected on campus, and as a senior describes: "We have an excellent Writing Center that really helps students improve their writing ability. We also have an excellent counseling center for students struggling with personal issues. Finally, Colorado College has a plethora of student organizations and other opportunities that help students to feel socially accepted and to make friends."

As a whole, CC is special because of the "Block Plan, which allows for travel, research, and concentration in one thing at a time. The school is special because it is like a little bubble in the Rockies; it is like a whole city that is tightly knit as a community." Says another student, "For students who want personal attention from their profs, who want to be challenged by a rigorous academic schedule, and who are interested in learning to appreciate learning, than CC is the right school for you."

What Happens After College

Students major in a wide range of disciplines that will lead them on to graduate school or directly into the workforce. The strength of the sciences, especially biology and geology, prepares many students for entrance to medical school and graduate school in the sciences. Sixty percent of each class enrolls in professional or academic graduate programs within five years of graduation, a figure that is comparable to that for the other top liberal arts colleges. A survey of Colorado alumni who went on to further study indicated that they felt far better prepared in their writing skills than a majority of their fellow graduate students, thanks to the emphasis on writing embedded in the undergraduate curriculum. Many students enter law and business schools after majoring in political science, history, and English.

DAVIDSON COLLEGE

P.O. Box 1719
Davidson, NC 28036

(800) 768-0380

www.davidson.edu

[
Number of undergraduates: 1,600
Total number of students: 1,600
Ratio of male to female: 51/49
Tuition and fees for 1999–2000: $22,228
% of students receiving financial aid: 34
% of students graduating within five years: 89
Endowment: $252,000,000
Endowment per student: $157,500
]

Overall Features

Davidson is still affiliated in an informal manner with the Presbyterian Church, which founded the college in 1837. This relationship is evident in the strong emphasis on values, ethics, and the honor code that are part and parcel of the academic and social guidelines prominent on campus. Its mission statement presents a clear picture of what a student and faculty member can expect when he or she becomes a member of this small, intimate community: "The primary purpose of Davidson College is to assist students in developing humane instincts and disciplined and creative minds for lives of leadership and service." The college "emphasizes those studies, disciplines, and activities that are mentally, spiritually, and physically liberating." This ethos is important to understand in considering Davidson as a place in which to study for four years. Given the college's significant commitment to educating its students, Davidson's faculty are evaluated for promotion both on the quality of their teaching and their research, which informs their interaction with students. Davidson has long enjoyed a reputation for close-knit relationships between teachers and students, small classes, and offering opportunities to engage in research projects and independent study with experts in students' fields of interest. The honor code is another defining element. The requirement to report a fellow student for any act of dishonesty is taken seriously by all members of the community. This contributes to the tone of openness and sharing that many individuals comment on in describing their experience. Davidson is a small, handsome campus that is a designated arboretum located in a small town. There are fraternities for men and eating clubs for women on campus, but all are nonresidential. Students therefore get to know virtually all of their classmates, since they all live in campus housing that is rated favorably by students. True to its founding and its location, Davidson is traditional in tone and student body. There is concern for community social action on the part of many students, but not too much interest in political activism, nor is there a great deal of racial, ethnic, or religious mix in the population. The two primary activities on campus are studying and athletics. Both are treated very seriously by a majority of men and women. The academic standards are demanding; faculty expect a high quality of work and active participation in the classroom. Most students plan to continue their education into graduate school. Seventy-five merit-based academic scholarships are awarded each year to attract students to this serious learning community. For so small a college, the quality of athletics is also of a high standard, with some twenty sports played at the Division I level. Many men and women participate in intercollegiate teams or do club sports. There is ample opportunity for any good high school athlete to join a team at some level of play. A number of athletic scholarships are offered for the very competitive athlete.

What the College Stands For

A senior administrator describes Davidson's educational goals, which echo those of many of the liberal arts colleges in the Hidden Ivies: "To provide the best preparation we can for lifelong learning; to prepare our students for lives of leadership and service; to provide a thorough education in the liberal arts and sciences; to teach students to think critically, to write well, to acquire a broad and thorough knowledge in a variety of disciplines." Another notes that the college aims "to produce thoughtful, service-oriented alumni who will be leaders in their communities, who, because of their learning experience at Davidson, are prepared to be critical problem-solvers with a thirst for knowledge." The college has increased its ethnic, geographic, and religious diversity over time, admitting women in 1974 and expanding its student population. It has worked to offer social options beyond "the traditional Greek system" and increased opportunities to study abroad. The college's honor code is said to establish "an atmosphere of trust" and to serve as "the bedrock of the learning experience." "In addition to providing an environment where students take self-scheduled exams," says one administrator, "and they don't talk about their exams to other students who haven't taken them yet, the honor code also pervades the whole campus in a more subtle way. Bikes and laptops don't get stolen. Money found on the sidewalk gets returned. The Davidson honor code helps develop in students, faculty, and staff a deeper sense of humanity."

Overall, we hear about the importance at Davidson of the college community and the personal relationships formed on campus: "There is an uncommon satisfaction with the Davidson experience that is obvious and pervasive in the students and alumni—enthusiasm and gratitude abound—primarily due to the care, personal attention, and quality received in all aspects of their experience here." Another dean says, "Our network of support—academic, administrative, and personal—helps students. From faculty advising to career counseling to personal counseling, we are here to help our students. Also, our students are quite well-motivated, reducing their needs."

In the future, Davidson hopes to increase the diversity of the faculty and the student body, to expand the opportunities for earning academic credit outside the classroom through such programs as service learning and internships, to continue to review its curriculum, to introduce "in a rational and planned manner more technology into the learning process," and to provide more financial aid, particularly for middle-income families. Currently, one-third of Davidson students are on some kind of scholarship, and 14 percent are non-white. Clearly, the college wants to improve those numbers.

Curriculum, Academic Life, and Unique Programs of Study

The academic requirements are rigorous and traditional in nature, what one administrator refers to as "an unwavering, strong core liberal arts curriculum." There are a number of distribution requirements, including a course in religion, which all students must fulfill. However, students comment on the easy access to their teachers if they need help and the support of their peers. This is not a community known for intensive competition among its students. For the very intellectually interested learner, a two-year humanities program based on the great books concept is available. Davidson has always been praised for its strong science programs and thus many

students have prepared successfully for medical school and other graduate degrees. Another program of note is the Medical Humanities Program. A new life sciences center will enhance the curriculum and teaching in this field. Many undergraduates major in history and government and political science as a foundation for attending law and business schools. One-half of all students participate in the many study-abroad programs the college sponsors. These include the Dean Rusk Program in International Studies based on campus and programs abroad in Mwandi, Zambia, India, Germany, France, and elsewhere.

The college offers a preorientation program for non-Caucasian students and hosts a number of multicultural and international clubs and organizations. A full-time dean works with these groups. The Black Alumni Association and the Second Family Program, which matches students with local host families, also help students to adjust to the college. Internship possibilities have expanded in Charlotte, a city twenty miles away that has experienced much recent development. The college also stresses its commitment to service programs and student-faculty research opportunities and interaction. Support services, including learning and counseling centers, resident advisors, and faculty advisors also help students to succeed.

Major Admissions Criteria

Davidson is looking for "bright, ambitious, open-minded, curious, honorable, physically and spiritually active, self-motivated students who are self-starters, willing to work hard, to live within a rigorous academic environment with high expectations and equally high results," to use the words of one institutional leader. Important criteria used in evaluating applicants are: "Ability to clearly articulate and present their thoughts in an essay; a strong record of achievement in their academic work; an extracurricular record that displays depth and meaning rather than breadth." Another administrator gives the major admissions criteria as "academic performance, extracurricular involvement, and personal recommendations." Another notes "high school academic success, rigor of high school preparation, and evidence of leadership and/or involvement."

The Ideal Student

The right student will enjoy an exceptional total experience in this college. Here is one of the finest environments for the more conservative, serious learner who enjoys active participation with others in the classroom, on the playing field, in the dormitory. Any young man or woman who wants to commit to the quality of life and spiritual values, not necessarily religious in nature, that Davidson stands for will enjoy college life and grow in many ways. Be prepared to work hard, make many intimate friends, and find numerous social and athletic outlets for your energy and enthusiasms. Do not expect to find a politicized campus tone, a free-wheeling curriculum, or a highly diverse student body.

"We're looking for students," says an administrator, "who are willing to devote time to extracurricular activities, who have a joie de vivre, and who are decent human beings. The students who seem to do the best here are the ones who recognize the academic and extracurricular opportunities Davidson offers and who take advantage of them." Another dean wants students who are "excited by ideas and the life of the mind," and who have a "willingness to be involved," are "anxious to put ideas into action," and have a "commitment to honor and integrity, leadership qualities, a keen intellect and disciplined work habits."

Student Perspectives on Their Experience

A member of the class of 2002 and a Bonner Scholar at Davidson had just returned from Bangkok. "Davidson sponsored my trip there for a month teaching English and working with a monk on a basic English-Thai audiotape for beginners. All in a day's work I guess . . . it was lots of fun." That says a lot about the kinds of things one can do, even in a small liberal arts college in North Carolina, and many students reflected these kinds of sentiments.

Students choose Davidson for the faculty, the location, the friendliness on campus, and the academic strength of the college. Says one current student, "People tell you, 'You'll know where you're supposed to be when you get that feeling—you'll just know. Back when I was a high school junior frantically filling out twelve applications to college, I really had no clue what 'that feeling' was . . . We visited Davidson at the end of our trip, arriving late for the two o'clock tour. What happened next, I think, most aptly characterizes Davidson. A regular student just stopping in to the admissions office introduced herself and offered to catch us up to the tour group, filling us in on what we missed. I thought 'Wow, that's a first.' And the information of the tour about Davidson's excellent teaching staff, academic reputation, and relaxed atmosphere (not cutthroat competition) among students just confirmed 'the feeling' I got when I was at Davidson. I chose Davidson over all the others for many reasons: the teaching staff, the close relationship between students and faculty, the absence of a graduate program . . . in short the academic atmosphere was ideal. And finally, just the openness of the student body and administration, the frequent, 'Hey, how's it going?', the fact that I can go to my prof's house for a movie and ice cream as part of class, that's what made the difference for me and why I chose Davidson."

Students like Davidson's educational style in and out of the classroom. "Davidson does a great job providing students the opportunity to take learning to a new level. By that I mean, Davidson is very much about educating the whole person not just in an academic sense, but as a well-rounded individual. Only so much can be learned from books—it's totally different to visit a historic sight, sit and talk with the homeless or a war veteran, go to Lake Norman and do your own research on the water quality. Davidson students, in general, take advantage of these opportunities by spending time abroad, volunteering in the community, and taking leadership positions on committees. Davidson caters to all of these needs through programs like the Dean Rusk Program, providing students with the opportunity for summer and semester study abroad. The Bonner Scholars Program, of which I'm a part, volunteering and serving in the local community in everything from tutoring to con-

struction, from coaching to free medical-clinic volunteering." A junior praises "teacher availability," explaining, "Teachers are always there for the students. All faculty office and home numbers are printed in the campus directory. Students are encouraged to contact teachers when they need help with something. A lot of teachers actually stay late. I remember going to the student union on a late-night snack break and being able to stop by my teacher's office and ask for help and also just chat about things that were happening." He praises the library ("Lots of volumes, the interlibrary loan system, a reserve for government documents, and the possibility of getting your hometown newspaper if you can get enough of your friends to fill out the required forms.") and the atmosphere ("Going to school in a small town is a great thing. From the chimes every fifteen minutes to getting your mail at the same post office as the professors and people in the town, there is a real homelike feel to the campus. Also, being only twenty-five minutes from Charlotte means that if you have to escape the suburbs you can—although most students seem to be content with staying at Davidson.').

The academics are challenging but fair, according to students. "Davidson professors do not believe in grade inflation at all," says one junior. "Students receive what they deserve based on their work during the semester. When you get an A, it means that you worked very hard and that it was exactly what you deserved. I know that when I graduate from Davidson my GPA may not be a 4.0 like friends of mine from high school, but I can tell you that every point on my GPA is something I worked hard for and in my mind will represent everything I learned while I was at Davidson. Personally, I would rather have an outstanding and rigorous education than be handed a 4.0 when I graduated." Another student points out the faculty support and attention that accompany academics: "The other great thing about Davidson is the way it somehow can present a rigorous liberal arts curriculum while still allowing for such a personal approach. I feel deep admiration and gratitude toward the faculty members for tailoring that curriculum to each and every student who walks through the doors. Coming to Davidson, I was required to take a math course. Math was never a strong suit of mine, and I hadn't received an A in a math course probably since sixth grade. I was scared out of my mind sitting down in Math 110 early in the second semester. Yet on the first day of class, before I had even introduced myself to the professor, Dr. King turned to me (knowing that I wrote about the school's basketball teams for the *Davidsonian*), and right off the bat he explained a concept of statistics using a scenario from the previous weekend's basketball game. The extra effort that Dr. King invested (by attending school sporting events, reading the paper, and making a concerted effort to incorporate that all into a lesson) made all the difference in the world to me and resulted in my receiving an A."

Continuing what seems like a theme among the small colleges: "The worst thing about Davidson in the eyes of the general public is our lack of diversity on campus. We're a Southern, white, upper-middle-class school. The Davidson administration/admissions office has been making a great effort to remedy the noticeable lack, especially of African Ameri-cans on campus." There is not necessarily a great deal of support from students for affirmative action. Yet they recognize the efforts of the college to diversify and appreciate the international and ethnic diversity that does exist on campus, noting that the school is welcoming and friendly to all students. Nonetheless, students "wish there was a greater degree of a social mix" and interaction among students.

Who succeeds at Davidson? "Any student (regardless of ethnicity, religious background, and so forth) who is excited about learning, is willing to give 110 percent in the classroom, is willing to be active in both the internal and the external communities, and is willing to take some risks will get the most out of a Davidson education." Again we hear about the social student who is also academically talented and interested. Another student notes: "Everyone who attends Davidson brings their different backgrounds and personal experiences. These experiences, when coupled with the curricula, help to expand on the subject matter within the class. I can recall a complex analysis course in which we had an exchange student from Germany. A great guy, he had already taken courses covering much of what we were learning in this class, yet he elected to retake this course so that he could get a different perspective on what he had already learned by absorbing the discussions in a U.S. college class versus one in Germany. While he felt that he benefited greatly from his experience in our class, I feel as though I gained much more from having him there as well. It is this type of educational environment that Davidson students create with the aid of the professor. A Davidson student will strive to make the most of the educational experience inside and outside the classroom."

"I find that my colleagues/peers are as helpful as my teachers," says one student. "At Davidson, the students are very willing to lend a helping hand. It's not a 'cutthroat, destroy your lab work' kind of a place. The workload is enormous—everybody knows it, both the teachers and students. Even though we gripe about it sometimes, it's something we pride ourselves on. We're all in the same boat and are willing to help each other make it through." Another agrees: "The overall feeling of camaraderie from the individuals, the dorms, even the lawns, aids to the education that one receives from Davidson. I mean, at Davidson students don't compete with each other, there isn't an 'I must do better than you' mentality on campus. Students here don't ask each other what their grades are, nor do they really care. The only person that a student competes with at Davidson is him/herself."

There is some criticism of the fraternities and eating clubs (sororities) on campus, with some students advocating their removal and others advocating coed houses. However, the opening of the Vail Commons as the open, coed, college-eating establishment and the student union, as well as the open-access, no-rush, and no-hazing policies in the houses, appear to have alleviated many student concerns. In addition to numerous social activities on campus and in Charlotte—including the all-school formal—students find other ways to bond with each other across many lines: "I love the community-building that exists on the freshman hall. It is an integral part of life at Davidson, and one that appar-

ently lasts beyond the first year on campus. Flickerball [7-on-7 touch football with unlimited forward passes, a sport played by all freshmen halls and by many upperclassmen as well] is a unique experience that forges lifelong friendships between hallmates."

Students reflect the administration's emphasis on and appreciation of the honor code, universally praising it in theory and in practice and in terms of its impact on their academic, social, and future development. "The HC governs every aspect of life at Davidson, from the classroom to the playing field, and it's taken very seriously. In a nutshell, it says Davidson students will not lie, cheat, or steal. In my choosing Davidson, the HC was very important because it grants the academic body, ironically enough, a great deal of flexibility. Teachers don't have to be in the exam room, I can schedule my own exams, I can take a test and have confidence that my studying will reflect on my test and no one else's, and so on. With the HC comes lots of worrying for prospective students . . . it worried me. Will I get kicked out? How strict is too strict? Does this mean I can't work with someone else? It's up to the teachers to what extent they wish to utilize the HC, but the point is, it's there and it works. Very rarely does anyone get expelled from college, but let it be said that there's some gray area that's gone over at orientation. What impresses me is that Davidson seeks people who value integrity in a world where integrity is seen less and less. Again, Davidson educates students in more ways than one. The bottom line is—it's there for the students."

It is worth including another student's detailed description of the honor code in practice, as it may help readers to gain a clearer picture of honor codes at Davidson and some other college campuses. "Nothing has been as powerful an influence on my life as the Davidson honor code. This is something that has become part of who I am as an individual. The honor code is not just a series of rules, it is in the heart and soul of every Davidson student, while they are at school and while they are away (even after graduating). The honor code basically says that students will not lie, cheat, or steal . . . Coming from a suburb of New York, I felt as though I had to keep an eye on my belongings at all times. But after locking my roommate out after his showers three or four times, we both came to an understanding that leaving the room open was not a dangerous activity. Heck, this last year I opted to purchase a Sony Playstation, and after other people on the hall learned about it, I had no problems allowing people to go into my room to use it when I wasn't around. That was the trust that I had in my peers and my trust in the honor code.

"Other situations, like academics, are equally affected. For example, my first test (we call them reviews) at Davidson was a Calculus II exam. I was coming from a school where students expected the teacher to remain in the room and make sure no one was cheating. Upon receiving the test at Davidson, the teacher asked us if there were any questions, and since there were none, he told us that he would be back five minutes prior to the end of class to wait for the return of the tests. Then he walked out of the class to do some work in his office. For a minute every student just kind of looked at each other and then we began to take our tests. It is this trust that teachers have in students, knowing that

they will not cheat, that allows us to take our tests without teachers looking over our shoulders. It also gives us the opportunity to take no-notes, time-limit, take-home exams. It is not uncommon for a teacher to give an athlete a take-home test to take on his/her bus if their traveling corresponds to a review date in the class. Also, when it comes to finals, students have the ability to self-schedule their exams. What this means is that over the course of the exams periods, students can take any of their exams in any of the exam periods that they choose. This self-scheduling is really what it means to have an honor code—teachers give us flexibility in arranging our exams, and students are trusted to not cheat while taking their tests. . . . Be it having the ability to leave your book bag outside of a classroom, knowing that it will be there when you return or just being able to talk to your teacher as a peer because you each have such a high amount of trust in each other, the honor code is something to live by. . . . On the way to a basketball game, a friend and I walked right over a twenty-dollar bill. Neither of us moved to pick it up, we just kept walking. On our way back we saw a note taped to the ground where the twenty had been; it said 'Twenty Dollars Found Here. If Yours, Please Call x1234.' That's the power of the honor code."

What Happens After College

After graduation, many Davidson students go to work in business organizations. Over time, however, more than 75 percent will earn a graduate degree, and a large proportion will enter the legal and medical fields. An institutional leader says, "Davidson students excel at positions that require creative solutions, leadership skills, and problem-solving abilities. They are good communicators, verbally and in writing. They can be given significant responsibility with little instruction. Although they may not have specific training for a particular job, they're smart enough and resourceful enough to be able to learn how to do a job exceptionally well."

The numbers show a broad array of postgraduate involvements, but Davidson's surveys of graduating seniors and those one year out of college show that about two-thirds of grads go to work after Davidson, with almost 40 percent going into business fields, 19 percent into education, 15 percent into science/medicine, 8 percent into non-profits, and 5 percent into law or politics. Of the one-fourth of students who go quickly into graduate school, almost one-third head into the arts and sciences, one-quarter into medicine, and 19 percent into law. And while students from the class of 1998, for example, were employed in twenty-nine states and twenty countries, eighty-two out of 260 students, the largest group, stayed in North Carolina to work after graduation.

EMORY UNIVERSITY

B. Jones Center
1380 South Oxford Road N.E.
Atlanta, GA 30322

(800) 727-6036

www.emory.edu

Number of undergraduates: 6,000
Total number of students: 11,000
Ratio of male to female: 44/56
Tuition and fees for 1999–2000: $23,130
% of students receiving financial aid: 42
% of students graduating within five years: 81
Endowment: $5,105,000,000
Endowment per student: $464,090

Overall Features

No college or university in America has changed more dramatically since its beginning than Emory. Founded by the Methodist Church in 1836 as a traditional Southern college to prepare young men of social position for the ministry, Emory has risen to the top group of the mid-size teaching/research universities, with a student body that includes as many Northerners and Midwesterners as Southerners today. The composition of the student body is fairly diverse as more students of color and all religious denominations have been actively encouraged to enroll. At present 25 percent of the undergraduate body represents various minority groups in America. Thanks to the infusion of hundreds of millions of dollars in gifts from friends and alumni in recent years, Emory is one of the wealthiest institutions of higher education in the nation. Its present endowment stands at more than $5 billion. As a result, outstanding faculty have been recruited from many of the other top universities, new programs of study and special research centers have been added, talented students have been offered generous scholarships based on merit as well as need, and the physical plant has become one of the most complete learning and residential centers of all colleges.

Emory fits a particular niche in the world of universities: It is a middle-size undergraduate college with major schools of business, law, medicine, and many academic disciplines available at the graduate level. The campus facilities are continually updated or new buildings erected to meet the needs of students. As an example, a $25 million state-of-the-art library was recently completed. New dormitories, a student center, and an athletic center have also been added in recent years. Emory's location on the outskirts of Atlanta and its size make it an attractive campus for strong students. Although the university guarantees housing to all undergraduates, only 65 percent choose to live in the dormitories. Many find the opportunity to live in an apartment in town or share a house with friends appealing. Over one-third of students join a fraternity or sorority, which provides an active social outlet for many students. Others take advantage of the numerous activities the university has created or spend free time in Atlanta. Emory is unique in the South for the fact that competitive sports are low-key, with none of the intensive recruiting and awarding of athletic scholarships that is very much the tradition in the region. There is no varsity football team, for example. There are nine intercollegiate teams for men and eight for women available to those who care to participate. There is a good deal of club and intramural athletic activity for students by contrast. Studies are treated seriously, as the majority of undergraduates have their eye on professional graduate school even as they enter the campus gates for the first time.

What the College Stands For

Emory sees itself clearly as a provider of a liberal education at the undergraduate level. As one administrator put it, Emory focuses on a "broad-based liberal arts education, with a comprehensive core curriculum." It puts a "strong emphasis on internationalism, languages, computer literacy. Our graduates become leaders across all disciplines. We want to prepare them to live in a diverse world, accommodate change, and to think,

communicate, and problem-solve effectively." Both Emory's undergraduate and graduate programs are highly ranked nationally, and the institution sees its rise in stature as beginning with a $120 million gift in Coca-Cola stock from a donor in the late 1970s (due to its Atlanta neighbor and benefactor's prominence, Emory is sometimes referred to as "Coca-Cola University"). "The growth in facilities, faculty, and programs has been meteoric," says one administrator. Contributing to this growth has been "unprecedented financial support, a favorable location in one of the country's most dynamic cities, superb facilities, the faculty and placement record, warm climate, and the growth overall of the sunbelt."

A dean lays out Emory's education goals: "Continue the search for knowledge as has been demonstrated by students during their formative education; integrate living and learning experiences into the educational process; develop and enhance leadership skills; ascertain that students have a broad-based liberal arts education, thus enabling them to have competencies in multi-disciplines; encourage continued educational pursuits beyond the undergraduate level, enhanced appreciation of the changing dynamics of the world, and enhanced appreciation for individual and group differences." To accomplish these goals, Emory provides "ample resources; numerous libraries; numerous computer labs around campus and in residence halls (residence halls are wired for computer/Internet access); small student-faculty ratio; the Freshmen Advising and Mentoring at Emory Program (FAME); specialized tutorials; individual and group academic tutoring through the Office of Multicultural Student Services; well-educated, experienced staff members."

Even with its size, Emory maintains a low teacher-student ratio, a strong residential college environment, and a broad Campus Life Program. "I would like to see the institution continue to stress values, morality, and ethical considerations in a complex world," says an institutional leader. "Emory has already done landmark work in these areas as they pertain to professional education in business, theology, law, public health, and medicine." Here we see the interconnections possible in a university, as opposed to a small college environment, where the graduate programs on campus provide resources and models for the undergraduate college.

One administrator notes Emory's forward-looking perspective: "Emory's best years remain in the future. How many schools can honestly say that? The school is still 'young' as national institutions go. We are still developing, growing, and experimenting. There is not the same aura as a place that is three hundred years old—there are new organizations, majors, programs. Student input is essential and sought."

Curriculum, Academic Life, and Unique Programs of Study

Keeping to its academic tradition, Emory requires completion of a core liberal arts curriculum. The requirements are fairly specific and take up about one-third of every student's program. Beyond that, the range and depth of subjects and majors to choose from is impressive. Students find the grading system rigor-

ous but the access to their teachers very easy. A large cohort majors in chemistry or biology because of their interest in premedical studies and the superb science departments. Others choose psychology, English, history, and business as preparation for graduate school. Students may double-major and pursue B.A./M.A. programs. Emory offers many study-abroad opportunities and a well-developed internship program. The university has unique relationships with the American Cancer Society, the Carter Center, the Centers for Disease Control, the Yerkes Primate Center, and the Emory Clinic and Hospital systems.

The "philosophy of 'athletics for all,'" says an administrator, "has some ninety percent of students playing intramural sports." This is complemented by the nationally ranked NCAA Division III athletic program. Emory competes in the University Athletic Association, which is modeled after the Ivy League athletic program and includes such universities as the University of Chicago, Washington University in St. Louis, Johns Hopkins, and New York University. The Emory campus is listed as one of the twenty-five "most wired" campuses and offers extensive use of the Internet. Professors and students, for example, can interact on "learnlink," a campus-based network. A key extracurricular program is Emory's national champion intercollegiate debate team.

"Volunteer Emory is one of the largest student volunteer groups in the nation," says one administrator, and students have representation and a voice from the Board of Trustees level to the Student Government Association. It provides many opportunities for undergraduate research across the curriculum.

Emory maintains a "need-blind" admissions program, providing a great deal of financial aid, as well as a national merit-based scholarship program called Emory Scholars. Emory has, according to one administrator, "extensive diversity. Twenty-five percent of the entering class each year represents ethnic minorities. The *Journal of Blacks in Higher Education* rated Emory as the number one school among the *U.S. News* top twenty-five for the number of black undergrads and faculty." Emory provides the Office of Multicultural Student Services, as well as many clubs, sororities, and fraternities catering to students' varying interests and backgrounds, and has the additional attraction of the diversity of the Atlanta area itself.

Major Admissions Criteria

Emory looks for "active students from high school, who are not only bright but student leaders in some way. Students comfortable with diversity and who seek ethnic, geographic, and religious diversity in their college experience. Students who enjoy being challenged and who are motivated to 'push the envelope' academically." Its major admissions criteria, according to an admissions office representative, are "curriculum chosen; outside time commitments beyond class; letters of recommendation from counselors, teachers, others."

The Ideal Student

Emory has great appeal for the serious student who is goal-oriented, even if he or she is not quite certain what that eventual goal

might be, who wants to experience a diverse campus life near a large metropolitan setting, who is willing to study hard, and who does not care greatly about a high level of campus spirit or spectator sports. Emory has great resources in all regards that are readily available to the mature, intellectually curious individual who will appreciate the solid foundation in learning the university provides. Opportunities to work at the elbow of a renowned teacher, writer, or scholar can be obtained but will call for individual initiative.

Student Perspectives on Their Experience

Emory students choose the university for its reputation and its focus on undergraduate education, but, also important, for its graduate and research resources as well as its urban location. "Emory was quite an enticing school," says one senior, "due to its wide variety of majors, activities, and resources to benefit from. Students have opportunities available at the Centers for Disease Control, the American Cancer Society National Headquarters, and the Yerkes Primate Center (all of which are walking distance from the campus and affiliated with Emory). As a student who has interests in the fields of medicine, anthropology, and business, I am always able to meet my needs and explore numerous options near the campus for gaining experience."

The university is said to do well at "introducing students to diverse cultures. Many students are introduced to cultures that they otherwise would have been unfamiliar with. Also, Emory has so many student organizations to choose from that it allows everyone to find their niche. The school does its best to provide for students with disabilities through the Office of Disability Services. They even provide note-taking services to help students who have a problem taking notes in class." The university "provides students with the ability to determine their own path. There are numerous options from which students may select the way they wish to get involved and grow."

"The students who do best," says a senior, "are those who immerse themselves in the culture of Emory. I truly believe that the students who do well in school are not those who always have the highest GPAs. Rather, the ones who are truly successful are those who have achieved a balance between scholastic achievement, personal development, organization involvement, and social growth. When you finish college, you should have grown as a person, not just academically." One drawback of Emory's diversity and multiplicity of options is its fragmentation, and thus poor communication across campus, according to one senior: "The multiple paths to self-determination and self-definition also serve to undermine the connectedness of different groups. Once people branch out they find much in common with others, but most are hesitant to do so." Thus, those who do best at Emory are "people who are willing to seek out all of the opportunities the school offers. Personality does not matter as much (whether you are an extrovert or shy). If a person is fundamentally concerned with learning and expanding their horizons, they will succeed."

Students at Emory are helped not only by their own drive, but also by the university's peer culture and support programs. "Personal motivation is the key factor that leads to success. Yet there are some services that help students to lead successful academic lives. The Office of Multicultural Programs and Services has a free tutorial program for challenging classes, and many students choose to tutor students on their own. There is also positive peer pressure to do well academically, and groups such as sororities and fraternities compete for scholastic awards." The honor code is felt to be important, as are the Hughes Science Initiatives Program for freshman minorities and women in science and the ESEP Program (Elementary Science in Education Partners Program), which allows students to teach science to elementary school students in the Atlanta Public Schools. "This gives students with expertise in science the opportunity to share it with children throughout the school system." A senior praises the "wonderful professors who care about students' development, helpful administrators, great facilities, an active student body, food, technology resources, and an overriding philosophy (shared by professors, deans, staff, alumni, and students) that a place of learning is only as strong as the commitment of all members of the community to that goal."

"Emory is full of professors and administrators who care about the students. This makes life more full and enriched and has made Emory a school that I am proud to call my own," says a senior. Another recalls "opportunities for student leadership in campus life; the openness of professors to instructing in all aspects of life; academic collegiality between all—these allowed me to develop intellectually and become more articulate and better at organizing group efforts."

Emory seems to succeed in its mission of providing a supportive, small-college educational program within its research university environment. "Emory is able to maintain a very personal and concerned atmosphere despite being a top research university. You truly feel a part of a community of learning instead of feeling you are being 'processed' through an academic system. The Emory family of alumni, trustees, faculty, staff, administrators, and students is a warm community that is willing to embrace each student in holistic development if the student is willing to reach out and contribute to the community."

What Happens After College

At some time after graduation, says one Emory administrator of the university's graduates, "about 70 percent enter graduate and professional schools. Emory has a tremendous reputation in medical and law school admission. The other 30 percent seek employment through our Career Planning and Placement Center. Emory is heavily recruited by most Fortune 500 companies, as well as education, government, and other agencies."

GRINNELL COLLEGE

P.O. Box 805
Grinnell, IA 50112

(800) 247-0113

www.grinnell.edu

[
Number of undergraduates: 1,300
Total number of students: 1,375
Ratio of male to female: 45/55
Tuition and fees for 1999–2000: $19,460
% of students receiving financial aid: 66
% of students graduating within five years: 85
Endowment: $1,020,000,000
Endowment per student: $741,818
]

Overall Features

Grinnell is something of an oddity due to its location: It is a small liberal arts college of national intellectual reputation located in a small town deep within America's bread basket. Its nearest academic neighbors are large state land-grant universities that serve the majority of in-state residents. Grinnell was founded in 1846 during the infancy of the settling of the Midwest on the model of the traditional New England independent liberal arts colleges. Renowned for its outstanding teaching and the quality of the student body, the college has developed over its long history a loyal following of generous alumni. This has helped to create the largest endowment in the nation among small colleges—over $1 billion. The ability to attract outstanding students regardless of financial need together with top faculty who want to focus their career

on teaching makes Grinnell a place where one of the truly top college educations can be had. Grinnell is one of a dwindling number of institutions that guarantees full financial support to all needy students. Each year the college gives out 300 merit scholarships alone. This suggests rather strongly the talent pool a serious learner will be engaged with for his or her four years. The environment on campus is open, intimate, intellectually liberal, and serious in tone. Students rave about the access they have to their professors and the opportunities to engage in research projects or independent study with them.

Grinnell has long enjoyed a first-level science curriculum that appeals to serious students who want to prepare for medical school or careers in teaching and research. Its graduates are perennially among the most successful in completing medical degrees and doctorates in the sciences. At the same time, many students are attracted to the humanities and writing programs that are also considered top-of-the-line. There are no fraternities or sororities on campus, and with a student body of only 1,300, this is an especially close community of learners and teachers. A new multi-million-dollar student social center provides for many student activities and brings many social and cultural activities to campus to compensate for the very limited off-campus offerings. Athletics are informal and treated casually by students. Here again Grinnell is an anomaly in the land of big-time intercollegiate football and basketball competition. Overall, Grinnell compares favorably with the very best of the residential liberal arts colleges in America.

What the College Stands For

Grinnell describes itself as "an undergraduate, four-year, coeducational residential college that seeks to develop in students analytical and imaginative thinking in the liberal arts. The college exists to serve students directly and society indirectly. The college's ultimate goal is to educate citizens and leaders for our republic and the world beyond our borders. To this end, Grinnell graduates should be equipped to pursue successful careers, satisfying personal lives, effective community service, and intellectually satisfying and physically active leisure." In 1998, Grinnell offered this presentation of the college's core values: "Excellence in education for students in the liberal arts, a diverse community, and social responsibility."

"Social responsibility and serving the community remains a value of our students . . . Independence of thought and rigorous academics remain the basis for a Grinnell education," says one administrator. "Grinnell faculty hold learning and teaching absolutely highest in their priorities. Students build very close relationships with professors. Grinnell is a place where a student works closely with faculty to set a course of study that will broaden perspectives and serve a student's future goals."

Grinnell's endowment has opened up many opportunities for the college and allowed it to carry forward its egalitarian heritage. As one dean notes, "The explosive growth of the college's endowment has enabled the college to concentrate on ways to achieve excellence, rather than on ways to balance the budget. This has resulted in both confidence and creativity among faculty, staff, and students in shaping the college's future. It has also allowed the college to remain need-blind in admissions . . . Our need-blind admissions policy and financial security have resulted in our having a very egalitarian atmosphere. All social, cultural, and academic events are free to everyone in the campus community. A significant amount of money has been set aside to support summer internships and research projects for students, so that these opportunities are not available only to those students who can afford them. Computers abound, so that students who cannot afford to purchase one have ample access." Another administrator says, "The financial accessibility of a Grinnell education reinforces the college's core values of equality of opportunity, respect for individuals, social justice, and community diversity."

Faculty, and their relationships as teachers, advisers, and mentors to students, are the keystone of Grinnell's education. As one dean puts it, "A caring, involved group of faculty are committed to teaching and student learning. Faculty members engage students in research in every academic division of the college. In addition to teaching, they support independent studies and guide readings, sponsor internships, and lead study-abroad programs. Faculty are incredibly invested in students. In turn, students are quite supportive of each other. Grinnell has a strong sense of community. The academic atmosphere, though rigorous, is non-competitive. Students do not compare grades. They are expected to work collaboratively in groups. They read and correct each other's work. This sense of community is enhanced by the absence of a Greek system on campus. There are few 'in' groups and 'out' groups among students at Grinnell."

Grinnell is a college that knows itself and what it stands for, and it knows the kinds of students who feel drawn to this small community. As one dean puts it, "I think that most underestimate what an advantage it is for us at Grinnell that students who come here come because they have selected Grinnell with intention and deliberation. Although the surrounding prairie is beautiful in its own way, students don't come for the atmosphere; they certainly don't come to ski; they don't come for the city life. They come because they find here a group of students and faculty who are stimulating to them, an atmosphere where they feel simultaneously at home and deeply challenged." As Grinnell looks to the future, with its strong endowment and growing national reputation, it is "positioning itself to be a leader in the liberal arts," according to one administrator. It is building on campus

and solidifying its extracurricular and co-curricular programs.

Curriculum, Academic Life, and Unique Programs of Study

Grinnell places a premium on writing and thinking skills, which are thus emphasized across the open curriculum that was instituted in 1971. There is only one specific academic course required of all students: the first-year tutorial. A student's tutorial professor is also his/her advisor. Faculty members design and teach over thirty varied topics, all of which emphasize intensive writing and discussion throughout the term. The college is also experimenting with a "senior capstone" elective program for upper-level students. Grinnell's great financial resources also have enabled the college to build a new science center, a learning/study library, and a fine and performing arts complex. These brand-new facilities mean students can gain a top education in the sciences, the arts, and the humanities. There is a impressive array of off-campus sponsored programs that students can choose from, especially the study-abroad centers. This appeals to a majority of Grinnellians both to satisfy their intellectual interests and to take a break from the small, rural campus. Grinnell has created a number of cooperative arrangements with larger universities' professional graduate schools that allow a student to complete the undergraduate and professional degree in engineering, business, law, medicine, and education over a shorter time than normal. Interactive learning is the engine that drives the Grinnell education. Students are expected to engage actively with their peers and professors in the classroom and to take a good deal of responsibility and initiative for learning in their field of interest.

Two key offices that represent Grinnell's service orientation are the recently established Office of Social Commitment and the Community Service Center. The college pursues a policy of student self-governance "which is now second-nature to the campus, and which colors policymaking in most areas of college life." A dean notes that "the open curriculum and the concept of self-governance work in tandem to create a distinctive atmosphere at Grinnell. On the academic side, the open curriculum presumes significant conversation between students and their advisors in crafting each student's academic program. This sort of freedom, flexibility, and engagement is mirrored on the student-life side of the college, where self-governance is at work. Self-governance presumes that students have the ability and the will to govern themselves, to agree on rules of residential living, and to respect the rights and property of others in the community. Residence life coordinators do not have responsibility for discipline and punishment, but are charged with community building." On-campus student resources include updated computing facilities and academic support from "learning labs" in writing, science and mathematics, and reading. Grinnell has a peer tutoring program and a New Science Project to help improve retention of students from groups that are underrepresented in the sciences.

Major Admissions Criteria

Grinnell seeks "individuals who are motivated, independent, intelligent, creative, and interested in being part of a vital and rigorous learning community," says one administrator. A senior dean looks for "students who have a thirst and passion for learning." Another says, "The college seeks high achievers who are genuinely committed to a liberal arts education and to a sense of community that encourages everyone's potential."

The Ideal Student

As several Grinnell undergraduates have commented, the decision of whether to choose this college for four years is a no-brainer. Grinnell offers no apologies for its size, location, and academic tone. It offers an exceptional set of human and physical resources for the intellectually curious and motivated learner who will embrace the advantages of a very intimate community. Those seeking a large university environment and style of education with all of the trappings that go with it will not find Grinnell appealing. There are no big-league sports teams to cheer for or intensive small college teams to participate in. Those who cannot imagine themselves living in a small-town, country setting are likely to suffer from claustrophobia quickly. A young man or woman who has thrived in high school on strong relationships with committed teachers who are passionate about their discipline and who take great pleasure in discussion and debate on academic and social issues will feel at home at Grinnell.

"The students who thrive at Grinnell," says a dean, "are those who are willing to be actively engaged in their education. Because students and advisors work together to build a student's four-year academic plan, students need to bring ideas, opinions, and interests to the table. They certainly do not need to be decided on a major when they begin. Most aren't. But they do need to be ready to bounce around a lot of ideas and to go after the things about which they feel passionately. In addition, Grinnell has proved to be a very welcoming place for students who may have been marginalized in high school because it was not 'cool' to be smart or different. Every fall it is exciting to watch students who, often for the first time, find others like themselves. Grinnell is known for its egalitarian, nonjudgmental atmosphere. Finally, Grinnell is a particularly good choice for students who have a commitment to social activism and service. The college encourages an ethic of service, and opportunities abound." Another administrator says: "Grinnell College has a tradition of respecting individual diversity. Campus life is characterized by thoughtful discussion, attempts to understand people whose experiences and choices are different from one's own, and receptivity to new ideas. Another strong tradition, dating back to the founding of the college, is the determination of generations of Grinnell graduates to put their ideals into action by addressing the central social and political questions of their time."

"We would like the 'secret' of Grinnell's success to become more open and visible," says an administrator. "Academically strong, creative, independent thinkers will thrive in this intense environment for learning. Grinnell's financial resources permit us to offer generous aid to students who are not sure they can afford a private college. This is a great college for gifted, adventurous students who seek to translate their ideals into action and to use their education to make a difference in the world."

Student Perspectives on Their Experience

Students choose Grinnell because of its professors, location, size, financial aid generosity (even for international students), and philosophy. "I wanted to go to school in a rural location," says one junior. "I was sick of cities. Also, I wanted to go to a school with very strong academics that would allow me to participate in cross-country and track. Grinnell is incredible at professor-student relations. You can approach any faculty member as their equal and most take time to talk to you about any issue. Also, the philosophy of self-governance followed by the college is a good 'real-life' reflective experience for all." Another junior says, "Grinnell's faculty is top-notch. Not only are they incredibly knowledgeable, their first priority is teaching. Their commitment to teaching makes the faculty Grinnell's most valuable asset."

"Students who are able to keep their academics in perspective do best," says one student. "Those who understand when it is time to buckle down with work, but also when the work does not need to be done. Highly motivated people do better, as do those who do not need a city to be entertained." A junior says, "Grinnell caters to the widest variety of students. I don't think a 'nerd' or a 'cool' person really exists because everyone has their own group of friends and can do well. Academically, the school is amazing. I've learned a lot!" A type of student who does well at Grinnell is "an open-minded liberal student dedicated to academics and to making a real contribution to campus life." Another junior adds that "although many different types of students succeed in differing ways at Grinnell, the school tends to reward those who take initiative. Students who are active academically or extracurricularly gain the most from attending Grinnell."

Students feel supported and connected. "There is a huge academic support network to which one can go for assistance in reading, writing, math, science, or academics in general. Also, your roommates, floormates, and classmates are always willing to lend a hand. Grinnell students are the best support network."

The campus is friendly at Grinnell, but diversity could be improved. Says a junior, "The college needs to improve upon its services for multicultural students. Campus life for them is improving, but the building of a new student center with multicultural space would be a huge plus." Some students report a decent amount of international and overall diversity on campus and a lack of a focus on socioeconomic distinctions among the students. Nevertheless, one student argues that "while Grinnell has committed itself to diversity, the vast majority of students still tend to be white and middle-class. Grinnell, despite its efforts, cannot seem to attract those of a lower class or racial/ethnic minority. While students are open-minded, open-mindedness cannot substitute for actual diversity."

For those worried about Grinnell's location, students who enjoy the college see it as a plus. "There is no problem finding enough to do in the middle of Iowa. Nothing beats a midwinter Iowa sunset." However, "the isolation of Grinnell is great for some, but suffocating for others. It's really difficult for some to get used to a very small town and small school in the middle of Iowa." Many students thus take advantage of Grinnell's strong on-campus and off-campus internship opportunities. As one notes, "The summer student-faculty research project helped me learn a lot and get some hands-on experience in working with equipment related to my field."

Grinnell is special because of "the people," says a junior. "There are people who are the most intelligent I have ever met, and who also love to have fun. It is academically rigorous, but not competitive. I love it! Grinnell encourages learning and thinking. I can't imagine making a better choice of a college to attend. I have gotten to continue dancing, be a part of student government, and be on numerous committees. I don't think this experience could happen anywhere else. I've met wonderful people and have worked with wonderful professors and administrators who have impacted me greatly. I will always love these four years."

What Happens After College

"We measure the success of Grinnell by the lives its alumni lead," says one administrator. "Grinnell prides itself on graduating students who become leaders in business—like Richard McGinn, '68, CEO of Lucent; leaders in music—Herbie Hancock, '60, jazz pianist; or leaders in their communities. Grinnell students come here with academic talent, independence, and creativity. They leave to lead successful lives and to give back." Grinnell students take many different routes upon graduation, including graduate academic and pro-fessional programs, and a diverse array of career paths. "Our first priority," says an administrator, "is to work with the students until they are able to clearly see the path they want to be on. Our career development office also offers a wide range of services to alumni . . . The commitment to servicing our graduates extends decades beyond graduation." Medicine and law programs are popular among grads, as is some kind of public service.

HAMILTON COLLEGE

198 College Hill Road
Clinton, NY 13323

(800) 843-2655/(315) 859-4421

www.hamilton.edu

[
Number of undergraduates: 1,700
Total number of students: 1,700
Ratio of male to female: 51/49
Tuition and fees for 1999–2000: $25,100
% of students receiving financial aid: 60
% of students graduating within five years: 88
Endowment: $306,222,000
Endowment per student: $180,130
]

Overall Features

A college with lots of tradition that was founded in upstate New York in the early 1800s to educate young men of that region in the liberal arts, Hamilton has undergone several significant changes in its recent past. Some 40 percent of students still come from New York State, but as Hamilton has gained an increasing national reputation for excellence, its student population is becoming more diverse. During the topsy-turvy 1960s when a number of new educational institutions were started, Hamilton, an all-male undergraduate college, assisted in the creation of the nearby Kirkland College for women. Kirkland was designed to offer a more progressive style of education with a strong bent in the arts and humanities. Its campus was built in a modernistic style across the road and through the woods from the hand-

some, traditionally designed Hamilton campus. The two colleges shared many programs and social events but there was clearly a difference in the nature of the student bodies and the modes of teaching. Over time it became self-evident that the most sensible plan was to merge the two schools and offer one curriculum taught by one faculty and managed by one administration.

Today, Hamilton College reflects the merger to some extent in tone and curriculum, but it remains a moderately traditional academic and social environment. The historic emphasis on presenting students with a liberal arts education directed by top-quality professors who are there to teach small classes and seminars remains true. A more recent change of import was the decision to make all fraternities and sororities nonresidential as part of a residential-life plan created by the college's trustees and administrators. The goal was to diminish the social role of the Greeks in campus life and their purported anti-intellectual force. All students now must reside in dormitories on either the original Kirkland facility or the Hamilton campus. Hamilton, like all of the small residential colleges of note, wants to put the emphasis on a more engaged, sharing community of young men and women, especially as it endeavors to enroll more students of color and other diverse backgrounds. The dormitories are being developed into clusters for social and academic activities. The administration is aware of the need to develop more social opportunities in this small-town, rural setting. Students have always praised the college for the friendliness of the faculty and the quality of its teaching. A significant portion of each graduating class will go on to graduate school, often having been influenced by a particular teacher/mentor.

What the College Stands For

"I want our graduates to leave this campus with an insatiable intellectual curiosity," says one top administrator, "with an unequaled ability to communicate thoughtfully, eloquently and compellingly; with the confidence, flexibility, and adaptiveness to meet unanticipated challenges; and a willingness to contribute actively to the societies in which they live." Hamilton College prides itself on the development of an "intellectual spark and social conscience" in their students, and measures its success in part by the significant involvement of graduates in the college. The key to Hamilton's success and approach lies with the faculty and with the college's attentiveness to students. An administrator says, "I think our

distinguishing feature is that everyone at Hamilton always puts students' needs first. Our faculty are all exceptionally talented teaching scholars, many with national reputations in their fields, but the real stars of this campus are the undergraduates. We have an extraordinarily able division of student life, whose deans, health and career counselors, and student activities professionals are devoted to making every student's experience as rich and satisfying as possible. Perhaps the most important reason students succeed here is because of other students. This is a supportive and caring community where competition is healthy, productive, and mutually gratifying."

Hamilton is seeking to expand its diversity, "not only socioeconomically and racially but increasingly ideologically, geographically, and culturally," says one administrator. The college uses a generous financial aid budget to help achieve this goal and support a number of multicultural organizations on campus. One administrator defines Hamilton's most important goal, critical thinking, this way: "This includes how to define an issue, to examine it from multiple perspectives, and to assess the merits of the various viewpoints." And outside the academic sphere, Hamilton students "are encouraged to participate in a variety of activities that provide them with exposure to different groups of individuals, as well as with the chance to make lasting contributions to the lives of others in the community."

Hamilton's faculty are committed to their students, and the residential community helps facilitate their engagement. "The fact that faculty are in close proximity to the students makes the formation of personal relationships between faculty and students possible," says a campus dean. Another administrator adds, "Hamilton students succeed largely because they themselves are a driven group of people, but the culture of this place is so focused on each student's success that they cannot help but do well. Obviously, faculty members play a key role here. Students cannot hide on this campus; it is a very connected place. As a result, a swimmer who is having difficulty in chemistry is surprised to learn that the coach and the professor not only know each other, but they have talked and have agreed to a plan to get the student back on track."

Curriculum, Academic Life, and Unique Programs of Study

Hamilton takes a strong stand on the skills it believes are essential to learning. Three writing-intensive courses are required for graduation in addition to the completion of an intensive independent research project in the student's major concentration. Departments such as physics, geology, history, English, and political science are particularly well-regarded. There is a commitment to maintaining small classes that are led by established professors for all four years. There are no teaching assistants at Hamilton. Numerous off-campus programs are made available, from studying abroad to interning in Washington, D.C. Students are able to complete a combined undergraduate-graduate degree course of study in law, medicine, and engineering with a number of prestigious professional graduate schools. For its size, the college offers a strong variety of arts courses.

One institutional leader points to three practices he sees as essential in a Hamilton education: "The first is a historic and ongoing commitment to the singular importance of writing and speaking substantively, cogently, and eloquently. We pursue this goal across our curriculum, through small classes where no one can be a passive learner. Our second distinctive practice, one which also crosses curricular and departmental boundaries, is our emphasis on a capstone senior experience for every student. Each Hamilton student is expected to write a thesis or pursue an intensive senior program that serves as the culminating experience of his/her four years. And the third practice, deeply ingrained in our college's heritage, is the expectation that students will always publicly demonstrate what they are learning through presentations, poster sessions, and performances." Another administrator comments that "while we no longer have a four-year public-speaking requirement, that ethos remains a part of a Hamilton education. The opportunity to present and then defend one's ideas before one's peers is now more broadly incorporated in our curriculum."

Some other key programs include the Higher Education Opportunity Program, a six-week summer enrichment program to boost achievement and retention of minority students; the Common Meal, which occurs daily and involves students, faculty, administrators, and staff; and the promotion of undergraduate research opportunities that join the practical and the theoretical, such as the summer research program allowing undergrads to work on projects with faculty. Says an administrator, "I have seen in my seven years on campus an emphasis on increasing opportunities for serious undergraduate research at Hamilton—and not just in the natural sciences. Students studying the impact of welfare reform on the local county government (and then presenting the results of that research to elected officials); students studying the impact of a proposed industrial park in a neighboring community (and, again, presenting it to local officials); a student studying the evangelical movement in Latin America (and then having a book published as a result of that research); and a student writing a novel as a senior fellowship project—all offer evidence of a pedagogical approach to undergraduate education that is increasingly focused on original research." Also of note is the Arthur Levitt Public Affairs Center, an undergraduate public policy think tank. The Levitt Scholars Program allows students to deliver their senior projects as oral presentations to high school seniors. Says an administrator, "For the Hamilton student, the experience is invaluable in developing confidence and understanding of one's subject matter, and it demonstrates again the college's commitment to oral and written communication. It also reinforces the notion that if you 'perform' an idea you own it."

Major Admissions Criteria

Hamilton is looking for "active learners, contributors to the intellectual, cultural, and athletic life of the college, whose passion about life and learning enriches the experience of everyone around them," says one administrator. In terms of important admissions criteria, "I believe that everything begins with academic excellence. We are looking for outstanding students whose performance in secondary school confirms their promise as lifelong learners. Through some combination of rank in class, standardized test scores, and counselor recommendations, I want to be convinced that a student's admission and matriculation will strengthen Hamilton's commitment to the pursuit of excellence. What gifts does an individual bring? The answers to that question drive our admissions decisions."

A dean says, "Besides outstanding academic records, we also expect our students to have engaged to a significant degree in extracurricular activities, including sports, community service, and cultural events. One predictor of success at Hamilton is a student's openness to new experiences. A second predictor is motivation to seek assistance, which all students, regardless of academic caliber, will need as they progress through the Hamilton curriculum." Another administrator says that "to the extent that it can be measured and revealed in teacher and guidance counselor recommendations, indicators that students will contribute and enrich our campus are very important. A student's high school record and extracurricular activities are also considered highly." A faculty member lists the three top criteria as:

1. Demonstrated work ethic.

2. Motivated to do something positive with his/her life.

3. Intellectually curious about the world.

The admissions office puts the top factors this way:

1. Academic preparation as measured by courses, grades, rank, and especially how much they have challenged themselves.

2. Standardized testing, giving us a perspective as we attract applicants from all over the world.

3. "Promise of contribution"—based on a student's interests.

The Ideal Student

Students who are happy and fulfilled at Hamilton like the friendly, traditional, and fairly homogeneous population and the strong relationships easily developed with faculty through the small classroom. They have little need for the amenities and attractions of urban life. A majority of students anticipate continuing their education in graduate school, especially in the professional fields. There are ample opportunities to become engaged through varsity sports teams, clubs and organizations, and student government. The collegian who enjoys a highly diverse community and a student body that focuses on political causes and social action will not find a large body of takers at Hamilton. An enjoyment of the out-of-doors helps greatly in making it through the long, cold, upstate New York winters. Anyone who expects stores and restaurants to be open all hours of the night will not be satisfied with the small town of Clinton.

Intellectually and socially, "The students who thrive here are those young men and women who are willing to take risks, who care deeply about giving something of themselves to the campus community, and who are unafraid of change and growth." A faculty member adds that "hard-working, curious, bright individuals are what we want and they do very well." An administrator says, "Those who do best here are students who want to be involved—in their classes and in the life of the college community." Hamilton is not a place for students who want to isolate themselves or get lost in the crowd.

Student Perspectives on Their Experience

Hamilton students are clear in their enthusiasm about the college and their reasons for selecting it. Says one senior, "I chose Hamilton . . . for three reasons: One. academic reputation and programs (at the time, I was looking for a school that offered creative writing); Two. dedicated faculty (I was blown away after I visited classes at Hamilton, one of which was an English seminar with *four* students and a professor—what attention!); Three. Frankly, I enjoyed myself at Hamilton more than at other schools. People were unpretentious, easygoing, and seemed to know how to enjoy themselves!" A classmate says, "Hamilton made me feel wanted from day one. The admission folks went out of their way to set up meetings with professors and coaches. When the time to decide finally did arrive, Hamilton made it clear how much they wanted me by offering a generous aid package. They never fell short throughout the whole admission process."

And clearly, the faculty-student connecting that goes on at the college is paramount for students. "Hamilton's greatest attribute, to my mind, is dedicated faculty. The college offers amazing opportunities to work closely with highly regarded and respected professors: whether in a very small senior seminar or class discussions continued over lunch or a celebratory dinner at a prof's house at the end of the semester. I never received anything less than spectacular attention and dedication from faculty members—it seems they truly enjoy teaching, and especially at a college that allows and *encourages* such close interaction between students and faculty. I consider many faculty members close friends, especially those in my department (English). . . . When I return as a seventy-year-old alum, I may not remember the name of this or that building, but I sure as hell will remember that life-changing conversation I had on its front steps with my advisor, mentor, and friend Professor Briggs. . . ." Another senior says that students are really helped by "professors, administrators, and resident advisors (I feel like once Hamilton lets you in, it does its best to support you and help you succeed)." "The people" make Hamilton special. "I know my professors and most of the administrators and many staff members on a personal basis. It really is like a big family community." And, "Hamilton makes it easy to find out what you are passionate about. The school not only helps students find their academic purpose and interest, but more often than not, lifelong passions."

One member of the class of 2002 testifies to this happening in his first year at the college: "Academically, Hamilton is extremely strong in the writing department. Teachers emphasize clear and effective writing in their classes. It was difficult to meet these high standards as a freshman, but over the course of the year I noticed my writing improve significantly. Hamilton also offers access to a writing center that many students take advantage of. Its main purpose is to help improve student writing through one-on-one discussion. There are so many different clubs and organizations on campus that it is easy to find a new and interesting activity to try or club to join. When I arrived on campus I was intending to row for the crew team. Instead I ended up trying activities I had never done before. I joined an a cappella group, having little musical experience (but enough to get in the group, I guess!). I also gave acting a shot by joining the student-run theater organization. Both experiences were an integral part of my freshman year. I never thought it would be so easy to join a club and jump right into something I'd never done before. Students are able to be part of so many activities due to Hamilton's small size."

"Right now, social life is Hamilton's biggest worry," says a recent graduate. "The college is in a period of transition—it seems it is trying to replace Greek life with something else, but I'm not sure anyone knows exactly what will take its place. From my first to my last year, I saw a significant decrease in all-campus frat and sorority parties, and a significant increase in 'Blockbuster nights' and drinking at the on-campus pub or one of the two dingy bars downtown." Other concerns expressed by students include a desire to be more involved in campus decision making, to increase all forms of diversity (although socioeconomic diversity, encouraged by Hamilton's financial aid program, is praised), and to have more communication between various campus groups.

The physical environment of the campus continues to be beautiful: "Hamilton's campus is very special and distinguished. Hamilton is *gorgeous*, especially in the fall, and its grounds are spacious and well-kept—we even have our own huge nature preserve, ranging from hiking/biking trails to a rare flora preserve to several college-owned reservoirs."

Who does best at Hamilton? Students seem to echo the thoughts of the administration on this count. "Students who create and take advantage of opportunities do best at Hamilton. There are plenty of people who sit back and give only fifty percent during their time here, but the ones who do the best, get the most out of the college, and contribute the most are the ones who give a hundred percent—or even a hundred and fifty percent! Astounding opportunities are available to Hamilton students—but not all of them display the hunger and drive necessary to realize the dreams Hamilton offers. Self-motivation helps people succeed, although students are sometimes 'pampered' a bit too much, in my opinion . . . And the courage to forge one's own distinct path is always a plus!" A recent graduate says, "Hamilton students are good at motivating each other to keep improving and doing their best. However, there is generally *not* cut-throat competition so there is a much more relaxed atmosphere." Another adds, "Students with energy, a love for learning, high academic endurance, and who are willing to take risks will do best at Hamilton. The expectations are high. Professors set high

standards and expect that they will be met. In a small classroom environment there is no hiding at Hamilton. You are forced by professors and other students to succeed."

Students strongly back Hamilton's study-abroad opportunities, which expose them to new experiences and return them to campus "to realize just what a privilege it is to be at Hamilton, a place where professors know the faces and names of all their students, a place where one can finish one's senior thesis and celebrate with a drink in the Little Pub afterward with faculty and friends . . . Study abroad also helps students broaden their perspective and bring *back* to Hamilton lessons learned elsewhere." Another admired program is Adirondack Adventure, a week-long, pre-orientation wilderness program for first-year students. Meant to ease the transition to college life, Adirondack Adventure often turns out to be a student's best week. The bonds of friendship formed on the wilderness trips often last all four years and more! The program lets students meet people, share anxieties and dreams, make friends with students *and* faculty (who often lead backpacking or canoeing trips themselves), and learn to put their trust in the community that will become their home for the next four years. "And best of all, Adirondack Adventure is fun! Even when, the year that I served as assistant director of the program, it rained constantly for three days while students were out in the mountains and lakes of Adirondack Park, they still came back *beaming* from their trips—with even more stories and bonds than if the weather had been gorgeous! There is truly something magical about the program."

For students interested in community service, HAVOC (Hamilton Action Volunteer Outreach Coalition) provides a good outlet. "This idea of service to the school and surrounding community," says a sophomore, "seems to be getting stronger with each incoming class."

Finally, students leave Hamilton ready for the challenges ahead of them and appreciative of their learning experience. As one recent graduate points out, "You will leave Hamilton with confidence and with the ability to express yourself and share your passions with others. Hamilton will make you work hard for this end result, but as I am already finding, it is well worth it."

What Happens After College

Hamilton graduates go on to a diverse array of careers and graduate programs, with over 60 percent in graduate school within five years. The college says, "Our students are prepared for everything" and notes law, diplomacy, health professions, teaching, advertising, and business as key areas of interest for alums. "We also claim an award-winning pastry chef, an internationally acclaimed ceramist, a former president of the Metropolitan Museum of Art, and numerous members of the foreign service with ambassadorial rank."

One dean notes that "perhaps the most important among the attributes we instill in students is the flexibility to respond to changing circumstances and interests as they travel along their own unique professional paths." Another administrator adds, "Over and over again, alumni tell us how Hamilton's three core values—writing, speaking, and critical thinking—have helped them outperform their professional peers." Hamilton's undergraduate research opportunities across many fields help students gain entry into top graduate programs and provide access to top jobs.

Alumni commitment and success are strong: "Hamilton is a small place with a big reach. In our peer group, we are among the colleges with the fewest number of alumni (14,500); in some cases, we have fifty percent fewer alumni. But they make up for it. They are passionate (in some cases, rabid) about this college. Both in dollars and volunteer time, they make their affection known. Typically, about fifty-five percent of all alumni contribute to the Annual Fund—an important measure of support, and one that ranks Hamilton among the top ten of all colleges in alumni support. The volunteer support is also amazing—from alumni willing to meet with seniors looking for jobs to admissions volunteers to fund-raisers, we have thousands of people wanting to be helpful . . . This is a very connected place."

HAVERFORD COLLEGE

370 Lancaster Avenue
Haverford, PA 19041-1392

(610) 896-1350

www.haverford.edu

[
Number of undergraduates: 1,147
Total number of students: 1,147
Ratio of male to female: 47/53
% of students receiving financial aid: 43
% of students graduating within five years: 91
Endowment: $240,400,000
Endowment per student: $209,590
]

Overall Features

In most of its qualities, Haverford resembles the greater number of small residential liberal arts colleges described in this book. The curricular emphasis is clearly dedicated to exposing students to a wide range of courses in the arts and sciences, with an underlying goal of enhancing their critical thinking and writing skills. As with the other colleges of excellence, a high premium is placed upon interaction between faculty and students. Classes are kept remarkably small and, consequently, the faculty expects a good deal of classroom preparedness and participation. A distinct advantage for Haverford students is that three-quarters of the faculty live on campus and make themselves available to students at all times. Haverfordians are intelligent, intellectually committed, and liberally minded. They are headed to profes-

sional careers, as are their peers at the other Hidden Ivies, but frequently with time out for travel, community service, or interning. The population of undergraduates is small, and all of them live on campus, fostering lots of interaction.

The ways in which the Haverford community takes on its unique personality have to do with its founding in 1833 as the first Quaker college in America and its special relationship with its neighbor, Bryn Mawr. While there are few birthright Quakers in the community today, the philosophy that underlies the Society of Friends is still much in evidence. There is a spirit of acceptance of all beliefs and racial and ethnic backgrounds; resolution of conflict over campus issues is arrived at by consensus and goodwill; a commitment to social action to assist the less fortunate is articulated in all areas of campus life; and the acceptance of the honor code as the active principle guiding academic and social life is acknowledged by all students.

Haverford has a generous financial aid program, which has led to a diverse population that includes 20 percent minority representation and over 40 percent aid recipients. Under the guiding force of the honor code, students take all examinations unproctored. They may even choose their own schedule and place to complete their exams. There is a required seminar on issues of social justice, and students gather together each term to review the honor code and school policies. If an individual violates the code in any way, he or she is required to apologize before the entire student body in addition to suffering the consequences of the student judicial committee. Consequently one can imagine the importance community plays on this campus. While students are concerned about social and political issues in the larger world, they tend to be moderate in their expression and actions. To serve others is more of the Quaker theme than political activism and pressure to change others' thinking. Another distinctive feature of the college is its unique and valuable relationship with Bryn Mawr. Since it has such a small enrollment, it is a great advantage for Haverford students to be able to take any course offered at Bryn Mawr toward their diploma. These are two separate colleges with their own individuality, and the relationship greatly enlarges the learning opportunities, not to mention the social advantages, for students. Those who wish to do so can even live on the other's campus and eat in the other's dining halls. Thus, this very small college doubles its resources for its students. A combination of the small size and the acade-

mic/social attitude of the college translates into a low-key athletic program that encourages participation by all who are interested. There are a fair number of intercollegiate teams, the most popular being soccer and lacrosse. There is no football program. There are no fraternities or sororities, so the residential focus is in the dormitories. Students tend to participate in the many extracurricular organizations on campus. Philadelphia is within easy reach by commuter train to allow time away from this most intimate of campuses when desired.

What the College Stands For

Haverford sees itself as having several unique attributes, among them: "*A critical balance*: Haverford has maintained what a former president described as 'the balance between intellectual rigor and moral judgment, of knowledge and its application to society.' *Quaker heritage*: What distinguishes Haverford from other highly selective colleges may be its Quaker heritage and thus, its egalitarian, inclusive community where individuals are important, all views are tolerated and there is a tradition of trust and responsibility among the students who attend. *A sense of community*: Most small colleges have a sense of community, but through its Quaker heritage and its 100-year-old honor code, a climate of respect and concern for others is central to Haverford's ethos. Students are made aware of the 'Haverford' community from the moment they decide to matriculate. In lieu of a required deposit, students are taken at their word when they indicate their decision to matriculate. *A conscious effort to remain small*: Haverford has purposely remained small, but it has also remained connected to the world at large and to the individuals who comprise the college community. In a world in which trends in education include distance learning, consolidated high schools and greater numbers of university satellite campuses, Haverford still believes that face-to-face interaction and collaboration are critical to the educational experience. *Fundamental commitment to a broad liberal arts education*: While higher education has undergone a number of pedagogical trends, Haverford has remained committed to educating young people in how to think critically, to understand fundamental theory and concepts, to weigh values and to form independent judgment."

Says the college, "Haverford provides its students with a liberal arts education in the broadest sense, to develop their capacities for thinking critically and comparatively. Students are expected to go beyond the mastery of facts and evolve into creative and innovative thinkers, experimenters or scholars. The college's rigorous academic program rests on the assumption that students who attend Haverford will become fully engaged in and contribute to the intellectual (and social) life of the campus. As a result of these goals, Haverford has developed an array of major programs that combine much of the disciplinary training of graduate programs with the kind of direct, imaginative and exploratory engagement appropriate to undergraduate education."

Haverford has developed and matured since the introduction of its coeducation twenty years ago and since it has fostered a national reputation. As one administrator notes, "The curriculum has been broadened significantly, the applicant pool and student body are now national, the faculty is larger and offers many more areas of academic inquiry. The social and athletic programs have increased significantly, the administration is highly professional, and the finances of the institution are strong."

The honor code is central to Haverford's mission and identity, and applicants are required to write an essay about how the honor code will relate to their educational experience. Says an administrator, "Haverford is the oldest institution of higher learning with Quaker roots in North America. While the college is nonsectarian, the intellectual and social life is very much influenced by the Friends' emphasis on spiritual insight, intellectual honesty, and leadership. Many of the college's more important programs and practices stem from its Quaker heritage. Students' academic and social life continue to be governed under an honor code established more than one hundred years ago. More of a philosophy than a code of conduct, the honor code means students are expected to develop a strong sense of individual responsibility as well as intellectual integrity, honesty, and genuine concern for others."

Curriculum, Academic Life, and Unique Programs of Study

True to its emphasis on intellectual exploration and independence, Haverford allows students to choose the majority of their curricula over four years. Thus, there is a great deal of flexibility and opportunity in the academic program, and the college sees over time an "increasing trend toward interdisciplinary, integrated learning." There is a language requirement regardless of one's major field of study. Although the majority of students concentrate in the social sciences and humanities, the college provides an excellent life sciences curriculum that many future doctors elect. Here is another of the quality liberal arts colleges with an outstanding record of preparing students for admission to medical schools. A majority of classes in all subjects place a good deal of emphasis on reading, writing, and presenting in a logical, systematic, and critical style. Faculty place much of the responsibility for mastering the subject on the student. And "the individual attention of faculty, administrators, coaches, advisors, and other students is key to students' succeeding at Haverford," says an administrator. Another adds, "The college is somewhat unusual in that the ties between student life and curricular areas are strong. The dean's office, for example, is structured to support both academic and personal advising. Because the deans work closely with faculty around individual student advising, there are few, if any, cases where students do not receive needed help. Besides the formal avenues for counseling students, there is a writing center and a math question center, there is career and pre-professional advising, and there are several programs for students of color and a network of student advisors who are trained to help classmates with a host of student life issues. An Office of Academic Resources helps to coordinate support for all students."

Some key aspects in the development of Haverford's academic programs are mentioned by one institutional leader: "Cooperation with Bryn Mawr College has matured to the point where students can enroll in individual courses, take complete majors, and reside at either institution. As the faculty has grown, so too has the breadth of the curriculum. There are many more instances of collaboration among faculty across various disciplines within the college and with faculty at neighboring institutions like Bryn Mawr and Swarthmore. Extensive renovations to a number of academic and residential facilities, construction of a campus center (and an integrated natural science facility) have occurred in the past decade, as has the installation of a number of technological improvements: an automated library system (a virtual library combining the collections at Haverford, Swarthmore, and Bryn Mawr Colleges) and a networked campus with instructional technologies in our labs and classrooms." Other facets of Haverford's personality and academic life again stem from its Quaker roots. "Consensus is the primary decision-making process used in all levels of governance among students, faculty, the administration, and board of managers," says an administrator. "Given the Quaker tradition against hierarchies, there is a minimum amount of titles and rank at the college. Few students refer to faculty members by their doctoral degree. As part of their requirements for graduation, all students must fulfill the social justice requirement by completing one of a number of courses available across the curriculum in which students examine issues of prejudice, inequality, and injustice."

While there is not a heavy atmosphere of competition for grades or graduate school admission, many Haverford students do plan to continue their education beyond the bachelor's degree some time in the future.

Major Admissions Criteria

Haverford's admissions office is clear about the three criteria they consider most important in the admissions process: "*Academic achievement and performance*: these include the strength of a student's secondary curriculum as well as grades. *Testing*: standard measurements like SAT or ACT; however, less emphasis is put on the importance of testing results in those cases where students have excelled in strong high school programs. *Contributions outside the classroom*: community service, other extracurricular activities, students who demonstrate 'well-rounded lives.'"

The Ideal Student

Haverford is a most attractive college for the student who thrives within a small, friendly community and enjoys an active learning and intellectual experience in small class discussions and research projects. People who care about others on and off the campus will find a home at Haverford. "Highly intellectual students who are interested in contributing to a small community are the kind who would do best and be happiest at Haverford," says one administrator. By contrast, anyone who wants to hide out in large lecture classes and take one or two exams a term to demonstrate mastery of the subject will not find Haverford College appealing. This is not the right college for the traditionalist who favors an atmosphere of conformity

and uniformity, nor is it right for those who are rabid athletic fans who follow nationally ranked teams. While there is easy access to a major urban center, the majority of Haverford students develop their social life within the small campus community.

According to one dean, "Being part of the community and giving something to others could involve participation in student government, the newspaper, or community volunteer work, for example. Because of the emphasis on student self-governance socially and academically, there is an expectation that everyone will become involved in some way or another. Close to seventy percent of the student body does volunteer work in some formal program. Seventy-five or more student organizations are run only by students. A very high percentage of students play sports." Others comment that, "When students refer to the social life at Haverford, they mean something more than going to parties—it's a sense of conscience." And, "Participation is just as important in academics where classes are small and student involvement is expected as well."

Haverford sees itself as a diverse and "unfailingly friendly place. Respect for the individual is implicit in all campus interactions and is one of the key tenets of the social honor code." Some 20 percent of Haverford's students are persons of color and close to 9 percent are students with international backgrounds. Of tenured and tenure-track faculty, 16 percent are of color. The college notes, "First-year students are made to feel welcomed and part of the community at the outset. A general orientation program called 'Customs' is designed and implemented by teams of students who serve as resource people and mentors for groups of students throughout their first year. Besides the mentors who live on the same dormitory floors with freshmen, there is a cadre of students trained to address social and academic issues who conduct weekly peer awareness workshops throughout the first semester on issues such as diversity. Prior to the start of their first year, students of color are invited to a program designed to help them identify and handle the challenges they will face at Haverford as persons of color. A winter institute is organized over winter break for all students and focuses on a specific issue of social justice. An Office of Multicultural Affairs is primarily responsible for supporting students of color academically, socially and emotionally as they progress toward their degrees. The office is also charged with developing and coordinating cross-cultural programming on campus. The college also has a number of 'interest-based' centers, lounges and houses set aside for diverse campus organizations."

Student Perspectives on Their Experience

Students choose Haverford because of its academic programs, its location near Philadelphia, its philosophy of education, and its student body. A senior from Delaware tells of her decision-making process, contrasting Haverford's environment to that of her state university: "When I was deciding which college to attend, my final choice came down to U of DE (which has a pre-vet program in which I was interested) and Haverford. I decided to attend Haverford for many reasons. While I felt confident that I wanted to pursue veterinary medicine, I also wanted the benefit of a liberal arts education. I ultimately decided that I would spend the rest of my life working with animals, but I would not get the opportunities that Haverford offered later in my career . . . I also like the small-school, tight-knit atmosphere at Haverford. After spending time on campus, I loved the feel of the college. Over the years, this feeling has become stronger and stronger. I honestly think of the Haverford community as a big family. While, of course, not everybody in the family gets along all of the time, there is a closeness here that I know I would not have found at a large state university. Not only am I close with the other students, but the community extends to include faculty and staff members as well. The professors go out of their way to get to know you. Dinners at professors' houses are a frequent occurrence and you often see students eating at the coop with professors or staff members. You get to know your classmates and the professors very well. I have taken many of the large classes offered on campus (Intro Psych, Organic Chemistry, Bio 200—the introductory bio course—Intro to the Study of Religion), but these classes are small compared to their counterparts at other schools. Once you get past the intro-level classes, the class size falls to an intimate level. This year, one of my biology courses had only eight people in it. I definitely would not have had this experience at a larger school."

Who does best at Haverford? A senior mentions, "You need to have an internal drive that causes you to want to learn, to work hard, and to be the best person you can. Haverford is an academically challenging college. You have to work long hours studying hard, fitting in your extracurricular activities as well as the social aspects of college. You need to be the type of person who lives the honor code, both academically and socially." Students are helped to succeed by their own initiative and drive. "With the honor code so prevalent in Haverford life there is very little outward competition to do well. No one tells you their grades if they don't want to, so as a student you have no idea how others are doing. In order to succeed you have be motivated by yourself, not by competition with others. Of course, the general atmosphere helps the general student population to succeed. Profs really care about you and know your name. You can talk to deans very easily. We have mostly self-scheduled exams; it is a trusting place, and conducive to learning and doing well."

The honor code is seen as a very important component of life at Haverford, in and out of the classroom. "This code is by far the most important part of this place, and I've noticed it affects my actions outside of the Haverford community as well." Another senior points out, "There are two main parts to the honor code. First, there is the academic honor code. This is what most people think of when they think of an honor code—no cheating, no plagiarizing, etc. We also have a social honor code. We are to respect each other and be respectful of each other. If we are having a problem with a student, faculty, or staff member, we should meet with them to talk things out. The social honor code is probably one of the most distinctive parts of Haverford. We are very committed to it. The honor code is very idealistic, and it is important to under-

stand that not everybody follows it all of the time (for instance, somebody may be loud late at night without really thinking about the fact that they could be being disrespectful to somebody), but everybody tries to follow it. All decisions are made by consensus. Haverford is not just a school, but just as important, it is a community. I honestly feel at home while I am on campus. I think of my classmates and professors as part of my extended family. I feel that the community is very close, which is partly due to its size and partly to its atmosphere. This closeness influences every part of our lives. The honor code can survive because of the community Haverford has created. Closeness leads to respect and this respect causes people to be respectful of each other. You do not lie to, steal from, or in any way hurt somebody whom you respect."

This is illustrated in the academic environment of the community as well. "The community aspects of the college help students to succeed. This is felt throughout the college. The professors are wonderful. They are always willing to meet with students and help them work through any difficulties they are having in class. My professors have always gone out of their way to help me. The community aspects of the college also lead to wonderful student-student relationships. We work together to reach a deeper understanding of subjects. Because of our commitment to the honor code, students are able to balance group studying with the individual work that is necessary. I feel comfortable going up to almost any person in my class to ask a question and I would take the time to answer the questions of any of my classmates. Professors take advantage of this group atmosphere—I have had an exam for which I was supposed to meet with a group to discuss all the questions before we submitted individual answers."

In terms of the college's downside, says a senior, "This place can feel very homogeneous. If you don't think or act in a very liberal way, it is easy for people to look down on you. People here are also exceedingly busy. Everyone is stretched too thin. Sports, singing, school. . . . Everyone gives one hundred percent at everything and it exhausts you and can stress you." Diversity overall at the college is seen as moderate, but the college is praised for its efforts to overhaul its programs in this area and to help students of different backgrounds adjust. Another senior comments about Haverford students that "they are generally kind, thoughtful, intellectual, and interesting people who have had some amazing life experiences and come from surprisingly different backgrounds, or at least have their own well-thought-out and strong opinions on all kinds of topics. I think that those who see Haverford as homogeneous aren't looking very carefully at those around them." And a junior points to such programs as Customs and Summer Tri-Co as helping to create "safe spaces" to address topics of concern. "Respectful, intelligent discourse is one of the most essential and wonderful parts of the 'community' atmosphere that is cultivated at Haverford." Another senior notes that "Haverford does not always show you how the 'real world' works. Sometimes people say that we are in our own 'Haver-world.' Out in the 'real world' people are not always nice, not always honest or respectful. People do not always trust you, nor do they always deserve your trust. I love Haverford and I would never give up the wonderful atmosphere that is present here, but you must remember that this is not the 'real world.'"

"It's tough," says a senior, speaking of the academic program. "It takes a lot of effort to really excel. You can have gobs of fun at college, but if you go here, also be prepared to work. It's safe. We are in a little suburb next to Philly, and although we are near a big city, we are really a tiny community, where everyone knows everyone else (this can also be overwhelming when you notice you know *everyone* on campus). I feel I should also mention Bryn Mawr, in order to get an accurate picture of Haverford. This institution, two miles away, provides a whole other college of classes to take. It is like being at a much bigger university because you can take classes there. It also, though, really distorts the male/female relationship on campus, which can be a bit odd."

Finally, Haverford students admire and respect the attention they get from faculty. As one junior puts it, "Haverford's major strength is the interaction that occurs between students and professors. Because it is a small college, Haverford's faculty members have the opportunity to really get to know their students, and vice-versa. The professors I have had the privilege to know and learn from truly seem to have the students' best interests at heart because they seem to have such a strong passion for what they do . . . Haverford is such a terrific school because with the 'external' competitiveness removed, students and faculty can appreciate learning in and of itself."

What Happens After College

One administrator commented that "a degree from Haverford has always meant the same thing: an excellent educational foundation that carries with it a notion of social responsibility and lifelong learning. As a result, our graduates are very interesting people, often with many dimensions to their lives, and they are very involved in their communities and their professions." Says another, "Haverford College graduates are prepared to go into a variety of fields after graduation." Information on where students went immediately after graduation in 1998, and on the career fields of members of the class of 1988, illustrate diversity, a high proportion of graduate/professional degree completion, and career preferences in education, business, health care, law, and communications. Over 40 percent of alums attend graduate/professional programs within five years of graduation, enrolling in many of the top programs in the country.

Class of 1998
Graduate/Professional School—19% total

Arts and sciences: 7%
Business: 1%
Law: 5%
Medicine: 6%

Employment—67% total

Arts/museum/architecture: 1%
Business: 15%
Communications: 5%
Education: 16%
Government: 3%
Health/medical: 2%
Community service: 6%
Law: 3%
Public policy/research: 6%
Science: 10%
Temporary work: 1%
Fellowships: 1%
Travel: 1%
Other: 1%
Seeking employment: 3%

Class of 1988
Employment

Arts: 4%
Business: 14%
Communications: 6%
Education: 22%
Health care: 14%
Law: 11%
Public Service: 4%
Social Services: 4%
Science/technology: 4%
Unknown: 16%

MIDDLEBURY COLLEGE

The Emma Willard House
Middlebury, VT 05753-6002

(802) 443-3000

www.middlebury.edu

[
Number of undergraduates: 2,100
Total number of students: 2,100
Ratio of male to female: 49/51
Comprehensive fee for 1999–2000: $31,600
(includes room and board, tuition, and fees)
% of students receiving financial aid: 40
% of students graduating within five years: 95
Endowment: $581,000,000
Endowment per student: $276,666
]

Overall Features

Middlebury has long enjoyed a reputation as a strong academic college in combination with a very upbeat social life and physically attractive environment. Its Vermont location and beautiful campus have made it an appealing place for generations of socially outgoing, athletic, intelligent young men and women who find such a combination of salubrious elements at the top of their wish list. Similar to its peer institutions, the small residential liberal arts colleges, Middlebury concentrates on delivering a strong foundation in the critical skills of thinking analytically and creatively, clearly expressing ideas orally and in writing, and appreciating the great ideas and issues of the past and present through exposure to the arts and sciences. The vehicle for achieving these goals has always been through the presence of top faculty who teach

and inspire students in the small-class situation. This has been the college's mission since its founding in 1800. An honor code is well entrenched in the academic life of the community. Exams are unproctored; students must take responsibility for their own behavior and report their peers for infractions of the code. A student-led judicial council hears cases of violations and renders judgment.

Middlebury is on a steady ascendancy in the selectivity of the student body, the caliber and breadth of its programs of study, and the physical resources for enhancing the academic, social, and physical well-being of its undergraduates. As impressive evidence of meeting its goals, the college's trustees have committed to a ten-year plan to build new facilities that will cost $200 million. These will include ski country–style dormitories, an Olympic-quality swimming pool, a new hockey rink, and a campus pub to feature all of the latest electronic communication tools. All of this is in addition to recently constructed academic facilities.

Middlebury's president has stated that the goal is to attract the very best of the high school talent pool and to compete directly with its closest rivals—Dartmouth, Williams, and Amherst. In order to help expand its academic and extracurricular offerings as well as to bring greater multicultural diversity to the campus, Middlebury plans to increase its enrollment at a gradual pace to 2,350 undergraduates. More faculty are being hired to maintain the traditional access by students to their instructors and the small-size classrooms. Other signs of change are the removal of the Greek system, replacing it with social houses based on students' interests, and the development of a commons system that brings students and faculty together in clusters to share interests and ideas. The hope is that the social fragmentation and anti-academic attitude of the old fraternity system will be replaced by a more intellectually positive ethos on campus. Many students are attracted to Middlebury both for its extensive off-campus programs of study and for its athletic and recreational options. Two-thirds of all students choose to study abroad during their four years. The college has long excelled in intercollegiate ski competition and Division III ice hockey, winning many championships in both sports recently. There are many varsity and club sports for the serious to casual male and female athlete. The Middlebury Ski Bowl is probably the best collegiate facility in the East, and the gymnasium and golf course get positive reviews from students. The vast majority of students take advantage of the location to ski and do other outdoor activities.

What the College Stands For

One institutional leader sets out Middlebury's goals: "We seek to educate students broadly in the liberal arts in the expectation that they will be leaders in whatever they may choose to do. We seek well-rounded students who will contribute to the college during their four years here, and contribute to society thereafter, and who will, not just through narrow technical expertise but through a range of skills developed here, lead lives of value to themselves and to society at large. We therefore view the residential, dining, social, athletic, and extracurricular components of our program as an integral part of a Middlebury education, which we define broadly, and which we believe will continue throughout a student's lifetime."

Middlebury sees itself as having a particular identity and continuity that is shared by students, faculty, the administration, and graduates. "In its essential, human features, the college has changed very little, and that is part of its appeal," says an administrator. "This is still a place where able, well-rounded students can come and be part of an active community that encourages them to broaden the limited reach of their own understanding. Our special claims continue to rest upon size and scale, the encompassability of this community, and the special type of education that results in such a place. This has less to do with enrollment numbers than with a commitment to preserve the essential spirit of the place. Nor is the nurturing of that spirit dependent upon permanent institutions or structures: Fraternities had their place, now they don't; we are decentralizing our student deans' offices and beginning a system of continuing residential communities with dining (the commons system). These may seem radical departures. In fact, they are done in order to preserve those things that make Middlebury—and Middlebury students and alumni—recognizable to one another over the years. So in one sense we have changed a great deal; in another we have changed hardly at all."

As Middlebury grows, the college will seek to maintain and continue to develop its residential, liberal arts community and identity, an aspect of itself that one institutional leader sees as crucial: "Keep in mind that, though we claim to be small and intimate, we are larger by far than most of the high schools attended by most of our students. To them we seem large, bureaucratic, impersonal. We need to do all we can to combat those tendencies. The creation of commons communities offers the best antidote. Moreover, the compelling claims of Middlebury will continue to rest upon the added value that comes from human interaction. Technology may broaden the definition of what it means to be a student . . . If we are to continue to be valued, [we] must offer education on a human scale, with human contact. This will be increasingly prized in the future."

Some of the things that make Middlebury special, according to one administrator, are: "Our location in Vermont. Our emphasis upon 'peaks of excellence,' which truly do define us and our special strengths: language, literature, international studies, environmental studies, our residential system, and opportunities for 'applied experiences.' The size and composition of our community, with its own brand of diversity. The small-town rural set-ting, the large number of international students, and Middlebury's outreach through Bread Loaf [the highly regarded summer English school and writers' conference] and the language schools. Not much doubt where you are when you're here."

Another dean notes that what helps students to succeed at Middlebury are "relationships with mentors of all kinds: faculty, staff, coaches, alumni, and townspeople. These relationships are facilitated by a size and scale that allow for multiple daily contacts with students both in and outside of the classroom. Everyone understands and supports the need for these relationships."

Curriculum, Academic Life, and Unique Programs of Study

Students must meet a set of general distribution requirements in keeping with the traditional liberal arts education. Specific requirements include the successful completion of two intensive writing seminars by the end of the sophomore year. The academic calendar includes a January term wherein all students engage in one course of study involving an in-depth project under the tutelage of a professor in the particular department of interest. Middlebury has a longstanding reputation for the excellence of its English, writing, history, philosophy, and language departments. More students major in these disciplines than is typical of most liberal arts colleges today, especially in foreign languages. The study-abroad programs are utilized by two-thirds of the student body, a percentage well above that of most of the top liberal arts colleges in the country.

For English and writing majors, the famed Bread Loaf summer program is readily available for advanced study and training. Language-immersion summer programs are also first-rate. As one administrator notes, Middlebury is "unique" because it is "a small liberal arts college that is also an international university—with summer programs in eight languages, schools abroad in five countries, and graduate programs in English literature in four locations." Another popular major is the interdisciplinary environmental studies program, which appeals to students who want to combine their love of the outdoors with concern for environmental problems. The new six-story, $47-million science center will make this field and science studies in general stronger than ever. Middlebury has a host of intellectual opportunities for the serious student that will continue to expand over time. As one administrator notes, discussing change and its causes at Middlebury, "the boundaries of knowledge have been altered; boundaries separating the traditional disciplines have shifted or crumbled. More than half our students now study abroad, now take double, joint, or interdisciplinary majors. Our students and their families expect certain services: career advising, counseling, every club or sport under the sun. None of these changes has been the result of a single act or decision by president or faculty. And yet these forces are prompting us to change. Technology is another example, prompting us from the outside to adapt to it, and, where possible, to exploit it on the inside. We do our best to choose how to respond to these various demands and to define a vision that embraces these choices and makes clear our priorities."

Major Admissions Criteria

Middlebury has joined the ranks of a growing number of outstanding liberal arts colleges that make submission of the SAT optional in consideration for admission. Applicants are required to submit three SAT II subject tests, however. This is in keeping with the admissions committee's emphasis on the high school transcript, meaning the strength and demands of the four-year course load, grades, teacher recommendations, and the applicant's personal application. Middlebury developed a creative approach several years ago to the dilemma of too many talented applicants for too few available spaces. It admits approximately 100 first-year candidates who will begin their studies in the winter term. This expands the admitted pool and fills the campus vacancies created by the large number of present undergraduates who leave for study-abroad programs.

Top admissions criteria listed by one dean: "The quality of the work done in high school is number one. The standard measurements of ability are number two, but the quality of the person is a must."

The Ideal Student

The student body has long been typecast as traditional-minded, white, upper-middle-class preppies who want a good education in a friendly, social, outdoors environment with good teaching available. There was no arguing the truth of this generalization in the past. However, efforts by the trustees and educational leaders of the college to broaden the base of the student body and the intellectual tone of the community is evident in the new buildings and academic programs, the expansion of the student body, and outreach on the part of the admissions team to attract more nontraditional men and women of talent. Some 40 percent of Middlebury students receive financial aid thanks to its healthy endowment and need-blind/full-need-met admissions, which are being used to diversify the community. The class entering the college in 1999, for example, contained about 15 percent U.S. students of color and 10 percent international students. Middlebury is a very appealing place for the smart high school graduate who desires immersion in a community of outgoing, physically active, moderate- to traditional-thinking students who take their education seriously but not obsessively. There is less engagement in community service or political activism here than in a number of the other campuses described in this book. As one dean notes, "Perhaps Middlebury's beautiful physical environment attracts students with an appreciation of nature, the outdoors, and adventure."

Student Perspectives on Their Experience

Students choose Middlebury for its academic programs, its location in the Vermont wilderness, its athletic programs, and its size. "The school excels, beyond all else I think, in offering a faculty and staff that are approachable," says a senior. "Professors, deans, custodians, security officers, advisors, and even trustees are all there to support and provide. They interact with the students and want to meet and engage us. This creates an atmos-

phere in which I found I was able to learn from people on many different levels, not just in the classroom." A junior says, "I think that in general Middlebury provides an amazing environment to learn in. Its small community feel heightens the college experience. Also, its internationally oriented academic program prepares students for participation in the current global workforce." Another junior adds, "The intimacy of the college is great—I've had dinners at professors' houses and I feel that teachers and deans are genuinely interested in helping us succeed. Classes are small and professors are encouraging. This school is just about perfect! We don't call it 'Club Midd' for nothing. . . . "

The students who do best at Middlebury are those who are involved in academics, but who look beyond the classroom as well. As a senior notes, "The same people who do well at other schools do well here: those who are focused on their work and what they want to get out of their education . . . The students who prosper and thrive at Middlebury are not always those who have the highest GPA, but those who have a very grounded perspective on themselves and where they are, who know what their interests are, and who take advantage of the rich resources—infrastructural, academic, social, and human." It is the interactions between people and through these resources that help students succeed. "The students who come here are smart and always deeper and more rich with experience than at first meets the eye. Learning from different strengths and using the infrastructural facilities is the greatest catalyst to success. If you're looking for more solid examples: interactions at the dining halls, dramatic improvements at the athletic complex, a Center for the Arts designed to encourage collaboration between music, theater, and dance, a new science center that incorporates the geography and psychology departments as well as physics, chemistry, bio, etc . . . ." A junior adds, "The person who does best is the one who finds the happy medium between academics and social life. I think what helps most of all is participating heavily in as much as possible. Grades help but I think in the end it's connections that open doors." A senior says those who do best are "students who are 'active' in all senses of the word! Those who are willing to work hard academically, who are excited to think, learn, and work with others, who like outdoor activities, who are excited to try new things. Those with a lot of energy and enthusiasm for life." A sophomore points to "the person who always expresses his mind and is willing to try new things. Middlebury helps this person develop in his own direction. Middlebury is like a guide that can point someone in a thousand directions; the student just has to experiment to find the right path. Ambition is the number one reason students succeed at Middlebury. It does not have to be ambition to do well academically, it simply has to be an ambition to follow what you want in life."

Again, faculty and classes get raves from students and form the core of their experience at the college. Says a junior, "My professors have all been wonderful, and I know that a small college like Middlebury can offer much more personal attention than a larger school. I took an Anthropology of Religion class with fourteen other students. Our professor shared stories of his trips around the world and led amazing hands-on activities that would have been impossible with a larger group. It honestly changed my life. Every class and every professor was outstanding, and it's great to be on a first-name basis with your teachers. When you're connected, it makes learning a lot more fun."

As Middlebury continues to change over time, particularly in its social and residential mission, it will confront its alumni and their views on what is right for "their college." The administration is well aware of the need for positive change and looking to the future. Students pick up on this as well. As one comments, "I would like to change the relationship between the college and some of the more disgruntled alums who see many of the recent changes as a threat to the college they once knew. This is a dilemma faced by every college when initiating change, but is truly a sign of how connected the alums feel to Middlebury." Another student, however, says, "Ironically enough the things I would change are the changes. I like the school the way it is and disagree with the direction it seems to be taking." Some feel the college is slow to change, slow to react to student concerns. And diversity at the college, while not extensive, is, according to many students, quite well developed in many respects and belies its public image to some extent. "Though Middlebury has a reputation of being one of the most homogeneous colleges, the diversity on campus is greater than is commonly perceived. The college is currently taking great steps to change that image and to increase cultural, ethnic, racial, socioeconomic, and other forms of diversity on campus. Predominantly the current conception of Middlebury College is a white kid with a Jeep Grand Cherokee from a New England prep school."

The commons system, one new aspect of campus life, was controversial when implemented, but at least one student is a strong supporter of its potential: "The commons has been very important because I see my voice making a difference. People believe one voice cannot do much, but I found that is wrong, it can change things. I helped create a more relaxed and respectful security system. I helped plan a dining hall, lounges, study rooms, a green, and underground parking in and around one of the current structures. I helped create the renovation ideas for an improved quality of life and a freshman dorm. Most of this was done through the commons and it has been important to me this year."

"Middlebury is a college that provides opportunity," says a senior. "A student who wants to do something can do it here. The environment is idyllic. The people are friendly, the natural surroundings sublime. You can get out of an education here what you put into it." Another senior says, "The school provides ample opportunities for any person's interests. It also admits students with a wide range of interests and strengths and the friendships that are formed are amazing. Middlebury provides a close-knit community that is felt and understood by everyone. Everyone seems to share the feeling of happiness and contentment, regardless of how stressful the academic load—which is very stressful. Somehow, the school provides an environment of relaxation. It also provides places where students are encouraged to hang out together, places for all students to enjoy regardless of interests." A recent graduate, and winner of a Watson Fellowship, says, "Everyone should know that at Middlebury students are enthusiastic and excited not only about their academics, but about their activities. Every student on campus is committed to some aspect of Middlebury. There are numerous chances to share, to help, to play, to learn, and to create."

What Happens After College

Says one administrator, "We keep close track of our graduates. We talk with them when we travel and when they return to campus. We ask about the ways in which Middlebury prepared them for the lives they now lead. And to a remarkable degree the responses are highly positive. Our students, in other words, are our students for life. To be sure, they have no way of knowing whether, had they decided to attend a different college, things might have turned out differently for them. They know only Middlebury. So comparative data can tell us only so much (and very little). But we do know what they think of Middlebury, and on the whole, their success and their attribution of that success offers us a good measure." A dean says that Middlebury graduates "seem to be prepared to do just about anything they want—and they do just about everything after leaving Middlebury. Increasingly, there are no geographic boundaries to what they decide to do, whether this involves work, study, or travel." A high percentage of alums will choose graduate school over time; and meanwhile will pursue over sixty different fields of endeavor within a year of graduation from the college. Law and education are two prominent areas of choice, as are English and foreign language and international arenas. Middlebury graduates do well in securing scholarships and awards, including three Watson Fellowships in 1999.

MOUNT HOLYOKE COLLEGE

College Street
South Hadley, MA 01075-1424

(413) 538-2023

www.mtholyoke.edu

Number of undergraduates: 1,850
Total number of students: 1,850
Ratio of male to female: 100% female
Tuition and fees for 1999–2000: $24,352
% of students receiving financial aid: 75
% of students graduating within five years: 82
Endowment: $327,000,000
Endowment per student: $176,756

Overall Features

Mount Holyoke, founded in 1837 as the first college dedicated to educating young women, has a long and honored history. It has stayed its course to deliver a liberal arts education of the first rank and foster in its students a strong social conscience. The classroom and the campus life are centered on the commitment to social service as a future career or through volunteer activities. This tradition helps explain the extraordinary generosity toward qualified students who cannot afford the price of the education Mount Holyoke offers. Three-quarters of the student body receives some form of financial assistance, truly an astounding ratio compared to that of most colleges in the nation. The direct consequence of this generosity is the diverse population of students of academic talent the college enrolls annually. The young women who

elect to attend Mount Holyoke tend to be serious learners who celebrate the idea of a woman's college where they can be leaders in and out of the classroom. They recognize that all of the resources of the institution are there for them to take advantage of, and this they do. True to the college's traditions, Holyoke women tend to be engaged in social causes and activism, be they political, gender, racial, or economic.

At the same time, a Holyoke student is more focused on her studies and her personal interests than on the traditional party life enjoyed at other colleges. There is plenty to do off-campus in this college-laden area. Some traditions do not fail: the men of Dartmouth, Yale, Amherst, and Williams still venture forth to the Holyoke campus for dates on the weekend. Since there is no history of sororities on campus, personal life centers on the dormitories that students help to run.

Quiet hours are elected by students in the residence halls to ease the academic workload. Faculty are respected for their expertise and ready availability to students. All classes are small in size and active participation is expected of all students. An honor code is in place and sets the tone for student attitudes. Exams can be scheduled individually and are self-proctored.

Mount Holyoke has a longstanding reputation for excellence in the sciences among liberal arts institutions. One-quarter of the student body majors in the life sciences such as biology and chemistry as a foundation for a profession in medicine, teaching, or research. The strength of the English department, however, makes it the most popular major on campus. The campus is very attractive and will appeal to the aesthetically traditional student. The classrooms, dormitories, arts, and athletic facilities are outstanding. Another historical tradition is the presence of excellent music groups—especially the various singing groups—and the music and theater programs. There is a superb equestrian facility and a golf course. But if any academic course or activity is not available on campus, a student can take advantage of the Five Colleges Group, which is comprised of Mount Holyoke, Smith, Amherst, Hampshire, and the University of Massachusetts. Mount Holyoke is also a member of the Twelve College Exchange, which enables students to spend a full year of study at one of the cooperating colleges.

What the College Stands For

The mission statement is short and clear: "Mount Holyoke College reaffirms its commitment to educating a diverse community of women at the highest level of academic excellence and to fostering the alliance of liberal arts education with purposeful engagement in the world." To implement this mission, the college has put forth a set of principles that are deeply reflective of the values of a liberal arts education and notions of serving the public good. The opening and closing paragraphs outline the highest of college ideals: "All human experience is education in some sense, and a liberal arts education does not exclude the sorts of learning that derive directly from the process of living itself. A liberal education differs from other varieties of education, however, because it places at its center the content of humane learning and the spirit of systematic, disinterested inquiry . . . Mount Holyoke is, specifically, a liberal arts college, concerned in a special way with the three ideas those three words bring together: freedom, learning, community of purpose. As a society of those who have come together for the sake of learning, the faculty and students profess and study those special arts of mind and spirit that they believe can free people— at least from ignorance, and perhaps from other poverties. The liberal arts are the arts of thought, perception, and judgment; the arts that foster humanity and civility of spirit; and it is these arts that Mount Holyoke College places at the center of its life."

Curriculum, Academic Life, and Unique Programs of Study

A reasonably flexible set of distribution requirements is in place to meet the college's aim of exposing students to the breadth and, in their major, the depth of the arts and sciences. Faculty are expected to do research and writing in their discipline to make them better teachers, and all faculty are required to teach. Mount Holyoke strongly encourages students to undertake independent study projects, research projects, or tutorials in a subject. Much of the success of the science majors in moving on to graduate programs comes from the opportunities to work closely with their professors on research projects. There is a wealth of opportunities for off-campus study programs sponsored by the college itself or in conjunction with the Five Colleges Group and the Twelve College Exchange. A January term is built into the academic calendar that allows students to undertake an internship or independent study in any field of interest.

Major Admissions Criteria

The college says "Mount Holyoke College seeks to enroll women who are outstanding in intellectual ability and personal qualities; differing in interests and talents; desirous of a liberal arts education; representative of many public and independent secondary schools; and with varied social, economic, and geographical backgrounds. The college looks for evidence of character, originality, and maturity, as well as sound academic training." The college is very individually focused in its admissions process, and candidates will be looked at personally and carefully. Key admissions criteria, according to the college: "Entrance is determined by the candidate's overall record and indications of promise as a student at Mount Holyoke. Greatest weight in a decision is almost always given to school

record and evaluations, but all information, including test results, special talents, particular goals, and evidence of determination, is studied carefully." Speaking for many colleges struggling with how to interpret and use the SAT and other standardized tests, the college says, "Generally speaking, tests are given little weight where there is evidence of high achievement and a rigorous high school program. When the board is uncertain about the level of achievement and the strength of the secondary school program, tests are usually given more serious consideration."

The Ideal Student

Mount Holyoke is dedicated to the task of training women for intellectual and social leadership in the larger community. Its environment is one of commitment to learning and active engagement in student government, the arts, and community service. There is life after studies, but the party animal category of student will not feel at home here. The casual student may find the academic requirements more than she really wants. The demanding Holyoke program is less about being an all-women's college than it is about the traditions of receiving a top education and putting it to use in a meaningful way. Holyoke alumnae are well known for their abilities and their active roles in the community. The network for assisting one another is impressive. The most satisfied Holyoke student is the young woman who has a genuine passion for learning and for exploration, who wants to participate energetically in campus organizations, and has a social conscience that leads her to engage in some cause. Those who have their sights set on graduate school as the next step to a professional career could not do better than to study at Mount Holyoke. A student willing to work hard in a more limited and low-key social environment than that available at many coed colleges, but in a place where she will be supported in every way, and a location where she will have access to ample cultural opportunities, but not those of a big city, will enjoy the enveloping safety and challenge of the Mount Holyoke College experience.

Student Perspectives on Their Experience

Women choose Mount Holyoke for its academic programs and reputation, sense of community, and most often because it is a women's college. Says a senior, "I looked at Mount Holyoke first of all because I knew of its excellent academic reputation, but then I started learning about the benefits of attending women's colleges and became more intrigued. I finally made my decision after visiting the campus for the weekend. I was inspired by the vibrant community of women and felt very comfortable." A recent graduate says, "Initially I wasn't interested in going to a women's college because I assumed it would limit my academic experience. Once I stepped on campus and met the dynamic women who live and learn here, I knew my assumptions were wrong and Mount Holyoke was the place for me."

The college is praised for its intellectual climate and ability to challenge and support students. A senior says, "Mount Holyoke is great at challenging every student to pursue their own interests, to think critically, and to make informed choices." A sophomore says that Holyoke "provides an endless amount of academic stimulation, support by students and faculty, and leadership roles and activities. Students are all women, so the role of women and their impact on society is highly stressed. The college places importance on the individual students." A recent graduate says, "Mount Holyoke produces intelligent, multitalented women with a strong sense of themselves and their place in the world." Another adds, "Mount Holyoke is a place where traditions are honored and progression is guaranteed."

Those who do best are "women who are intellectually curious and motivated, women who are open-minded, eager to try new things, women who are willing to share and learn from each other." Another says they are "strong, confident students who have done exceptionally well in high school, women who are overachievers, who are talented in multiple aspects." Others point to "anyone with an open mind, and the hunger to be challenged every day in every aspect of campus life"; "women who are willing to work and believe and continually challenge themselves and the world"; and "students who can work independently and balance their activities, sports, and social life with academics."

Student complaints, as might be expected, center around social-life options on campus and meeting men. Other areas of concern are residential availability and options. Nevertheless, one student argues, "I found that the party scene got boring rather quickly. Thank God I was at MHC where there are dozens of other options! Students can attend concerts and dance and theater performances, special lectures, film, poetry readings, and festivals as well as large and small parties." Students encourage prospective applicants to be balanced in their pursuits, and not to try to be "superwoman" by doing too many things at too high a degree of intensity.

The honor code is praised for setting the tone of campus life and creating flexible options for students in taking exams, and international and overall diversity at the college is seen as strong and positive. "The school does an excellent job of creating a diverse, supportive, comfortable environment," says a member of the class of 2002. "They also look to tomorrow and prepare you for the future. Mount Holyoke can best be summed up with the saying we are 'uncommon women on common ground.' MHC has every country, every state, every religion, and every socioeconomic background within our gates. And we take a lot of pride in this fact. We really try to celebrate diversity, not shy from it." Programs of note include the innovative Speaking, Arguing, and Writing Program.

Students feel empowered and connected at Mount Holyoke, with and by each other, the faculty, and alumnae. "Teachers at Mount Holyoke," says a 1999 graduate, "are fully invested in their students. When they see talent in a student they seek to develop and challenge it." A sophomore says, "What makes Mount Holyoke special is definitely the women who are here. I have been so amazed by the brilliance, talent, ambitions, and dreams of my peers." Another adds, "Mount Holyoke gives students a sense of individuality. All of the teachers I've ever had, even those in a seminar class, still recognize me, at least by face, and many even by name. It imparts a feeling of security that I haven't seen elsewhere. Mount Holyoke, for the first time in many women's lives, enables women to stand up as individuals and speak their mind, accomplish amazing feats, and excel in all areas of their lives."

What Happens After College

About 30 percent of Holyoke's graduating class will advance directly to graduate or professional school. Over time, about half of the 70 percent of students who head to work first will seek an advanced degree. Mount Holyoke sponsors an AlumNet program with 18,000 alumnae career mentors for students. Participation in summer and January internships help Holyoke students develop career interests and skills. Students win a number of major awards and fellowships, and have access to a good deal of active recruiting on campus and in Boston and New York. Key areas of interest and success for Holyoke graduates include such service-oriented areas as education, social work, public service, the law, and nonprofit organizations. The college notes that it "has produced more women Ph.D.s in the sciences than any other institution in the United States. Our alumnae are found in the capacity of research positions, as engineers, industrial chemists, physicians, astronomers, and software designers. At the same time, our focus on the arts results in many alumnae going on to become performers, directors, or playwrights, such as the Pulitzer prize winner Wendy Wasserstein." Alumnae also pursue careers in business, management consulting, investment banking, and insurance. The college sees its large number of international students on campus and the high participation in study-abroad programs as contributing to graduates interested in and prepared for internationally oriented careers.

UNIVERSITY OF NOTRE DAME

1 Grace Hall
Notre Dame, IN 46556-5602

(219) 631-7505

www.nd.edu

[
Number of undergraduates: 7,800
Total number of students: 10,200
Ratio of male to female: 55/45
Tuition and fees for 1999–2000: $22,344
% of students receiving financial aid: 46
% of students graduating within five years: 93
Endowment: $1,766,000,000
Endowment per student: $173,137
]

Overall Features

What high school student is unfamiliar with Notre Dame's fame due to its great football teams and overall excellence in major intercollegiate athletics? Its athletic prowess and tradition has made this mid-size university located in a small Midwestern town a known presence across the country. At the same time, how many future young scholars know very much about the outstanding educational programs and community traditions that make Notre Dame a first-rate university? Thanks to its loyal following of alumni, it has emerged from its founding in 1842 by the Order of the Holy Cross as a tiny, struggling regional liberal arts college into one of the most well-endowed research and professional institutions in the nation. Today, Notre Dame is a full-blown university comprised of undergraduate colleges of liberal arts and letters,

sciences, engineering, and business management. Its graduate schools of business and law have attained a national reputation in recent years, as have some of its doctoral departments. While its undergraduate colleges are more career-oriented than the typical small liberal arts school, there is a strong tradition of ensuring that students develop the skills that come from exposure to a broad range of arts and sciences.

In several ways, Notre Dame is a throwback to the more traditional nature of colleges prior to the upheavals the sixties wrought on most campuses. The student body is conservative in its political, social, and religious views. The vast majority are career-oriented, white, middle-class young men and women. An insight into the nature of the student body is the fact that the largest enrollments are in the colleges of business and engineering. This is one of a dwindling group of colleges that holds to parietal rules in the dormitories and on campus. All dormitories are single-sex and students must leave the rooms of the opposite sex by midnight during weeknights and 2:00 A.M. on weekends. Priests and nuns live in every dormitory and, while alcohol is allowed in the rooms, sexual activity is not, leading to severe disciplinary action if found out. There is a chapel in every dormitory for students to attend mass each morning. There are no fraternities or sororities on campus, so the dormitories are the centers for social life, particularly for students not involved in the hugely popular intercollegiate or intramural athletic programs. The isolated location of the university makes the campus a strong community and the focal point of the college experience.

Notre Dame is clearly a Catholic-centered university. Eighty-five percent of the students identify with the Catholic religion. A course in theology is required of all undergraduates. As it aspires to national prominence as an intellectual community, Notre Dame must contend with the natural state of tension that exists between the intellectual independence, areas of research, and teaching that define a great research institution and the strong Catholic traditions and philosophy that can restrain open-ended inquiry and research. There appears to be somewhat less control at present by the Order of the Holy Cross and an orthodox religious philosophy. There are fewer Catholic faculty today, which expands the points of view espoused in the classroom. Nevertheless, there is a vibrant and committed community of Catholic leaders and teachers who direct this growing university. True to its history of welcoming any student who is serious about his or her education, the univer-

sity has a very generous financial aid program in place. Almost one-half of those enrolled receive aid to make it possible for them to attend. A good number of academic scholarships are available as are athletic scholarships across all the sports teams. The spirit of the campus, the quality of the faculty, the great physical resources for learning and extracurricular outlets, and the strong counseling and advising support system results in one of the highest retention and graduation rates among all the top universities. This prairie college has come a long way from its earlier days of anonymity until a poor Norwegian lad by the name of Knute Rockne—who attended on a full scholarship—took over the reins of the football program and put the college on the map.

What the College Stands For

As one administrator recently noted in an address to the Notre Dame faculty, "The foremost goal of a university is to have great learning in the classrooms and on campus and to have brilliant publications emerging from the halls of our faculty offices. Whatever administrative initiatives we may undertake are minimal compared to this daily activity of learning and scholarship." Notre Dame's focus on combining undergraduate teaching and research scholarship showcases the university's identification of itself as a national research institution. The university's identity "has three major facets: Notre Dame is a residential liberal arts college, a dynamic research university, and a Catholic institution of international standing. Much of what is great about Notre Dame derives from the intersection of these three factors."

The dean goes on to spell out his view of the centrality of Notre Dame's religious values to the university's educational mission: "Notre Dame's spirituality in general and its Catholicism in particular enrich the liberal arts experience, with its ideal of educating the whole person. Prayer and liturgy belong to our students' college experience. Spiritual questions arise in all our disciplines. Perhaps only in a religious setting, where reflection on God, or, metaphysically stated, the absolute, is prevalent do we address life's most fundamental questions, which are increasingly bracketed at non-religious liberal arts colleges. Religion also brings to the liberal arts ideal a strong existential component. At Notre Dame learning and morality, knowledge and virtue overlap. Students on our campus pursue theology not as the disinterested science of religious phenomena but as faith-seeking understanding. Our students study history and the classics in order to learn not simply *about* the past, but also *from* the past. They read literature and are exposed to the arts because of their moral value, in the broadest sense of the phrase, and because we believe that beauty is the sensuous presentation of truth, not idle and meaningless play. Students employ the quantitative tools of the social sciences not simply as a formal exercise with mathematical models but in order to develop sophisticated responses to pressing social issues. In a world in which scholarship has often become antiquarian, disenchanted, and even cynical, at Notre Dame our work is shaped by our values and our existential aspirations. The generous commitment of the Holy Cross religious order to our campus and the ways in which they serve as role models reinforce this integration of learning and character, of college and community, of faith and life." As another dean puts it, Notre Dame's central mission is "to provide a high-quality education for the Catholic leaders of tomorrow."

Notre Dame sees itself as having a dual identity as well in its position as a "modern research university" with an interdisciplinary emphasis. The university encourages a focus on community, engaging in discussion, and developing "virtues." Notre Dame recognizes the positives of its strong athletic (particularly football) program, such as increased alumni involvement, support, and financial aid availability, as well as the negatives, including poor press around recruiting scandals, some students on campus more interested in sports than intellectual pursuits, and non-athletic students possibly feeling left out. These are areas that the university is working to improve. Other areas of focus include emphasizing academics at the university and improving Notre Dame's reputation and recognition in that area, targeting resources toward stronger faculty hires and student recruitment, increasing diversity among students and faculty, and developing academic strengths across the undergraduate and graduate curriculum in addition to the traditionally strong departments of theology, philosophy, and medieval studies, for example.

Curriculum, Academic Life, and Unique Programs of Study

Notre Dame's academic program is somewhat unique in its curricular structure and requirements. All first-year students, irrespective of their future field of concentration, are enrolled in the College Studies Program, which has its own dean and faculty. They must complete a ten-course program of study that includes an interdisciplinary seminar and a grammar/composition workshop. Second-year students enroll in one of the four specialized colleges of choice: arts and letters, business administration, engineering, or science, where they will have to fulfill specific course requirements. Unlike the strictly undergraduate colleges of note, Notre Dame does rely on graduate students in the arts and sciences to teach or lead discussion groups in undergraduate courses. Here they are similar to the larger university model. Overall, the curriculum is structured and content-oriented in contrast to the more flexible options of many other colleges.

Notre Dame is exploring ways "to help bridge academic and residential life, including offering selected classes in the dormitories, increasing faculty-student contact in the dining halls, and giving students greater responsibility for cultivating intellectual life." The university has begun a pilot program in peer tutoring and has emphasized improvement in "advising, including the informal mentoring of students." A dean notes that "this fall the College Council passed a policy for departmental honors that will encourage more ambitious and enriching intellectual experiences for undergraduates, including the writing of more senior theses. Our newly created Undergraduate Studies Committee has as one of its obligations the fostering of greater extracurricular learning opportunities. The Arts and Letters Student Advisory Council was revived last year and has proposed several ideas for enhancing student-faculty interaction and improving services to students."

Discussing Core, one of the university's central educational requirements, this dean reports, "The college will fund groups of five or more faculty members from at least three departments and two divisions to design new sections of Core. Topics such as poverty, justice, ecology, and classic works from diverse disciplines have been mentioned as possibilities. A course on poverty, for example, might include biblical and theological readings, philosophical works, social scientific texts, and art works. During this transition period experimental sections of Core would be offered, even as the traditional model continues. In other words, students may continue to take the traditional Core, which focuses on God, self, society, and nature, or select any one of the new versions of Core. The thrust of the proposal is to invite faculty members to stretch beyond their disciplines, but with topics that are particularly inviting to them and which may resonate with their research and development interests. Much like the concentrations that have developed in the college in recent years, this new option for Core should create clusters of faculty members working together on interdisciplinary topics that address value questions with students."

Major Admissions Criteria

Notre Dame has become increasingly selective, with about 35 percent of 10,000 applicants admitted for 1999. The university sought and found a bright applicant pool, averaging in the top 5 percent in their class and a 1,358 SAT. Of note is the fact that 83 percent of admitted students were involved in community service. As one dean says, top admissions criteria are "academic course load, grades in those courses, and test scores."

The Ideal Student

It should be self-evident that a prospective candidate needs to embrace the special traditions that identify Notre Dame. While it is not necessary to adhere to the Catholic faith, it is important to respect Christian values and teachings. It is also important to enjoy the sports-oriented, conservative student body whose long-term goals are to find success in a traditional career. The social or political activist and the free-spirited individual who prefers a more open curricular style of learning will not be happy, nor will the student who wants to experience an urban life for four years.

Students should expect and prefer a strong sense of community and college identity, as well as close personal support and supervision. As one dean notes, students succeed in part because of "an infrastructure that is so supportive of personal success. The halls are the 'backbone' of the campus, with eighty-four percent of the students living on campus. There is no Greek system. Full-time counselors/professionals are in the halls and in other units on campus (First Year of Studies Office, Campus Ministry Office, Student Counseling Center, etc.) . . . Notre Dame is a highly selective, residential academic community within a faith tradition."

Student Perspectives on Their Experience

Students often have trouble distinguishing and choosing between Notre Dame and two of its top rivals for attention, Georgetown and Boston College. One student who was admitted to all three described her choice this way: "The one thing that I saw as a negative about Notre Dame was location. Part of what was so appealing about the other schools was that they were set in the middle of great and exciting cities and there could never be a shortage of interesting things to see and do. I have come to realize, though, that the campus of Notre Dame is one of its greatest assets. It is exactly how I envisioned a college campus and one of the things that I love about Notre Dame is that it is all-encompassing. People often speak about the Notre Dame family, and there is really no better way to characterize life here. It is a community of people who live and learn together and it affords an opportunity for people to develop relationships unlike any others and to share unconditionally. I have friends that I not only study with, but eat, sleep, and pray with, too. Notre Dame puts things into perspective and brings great balance to what can be an extremely hectic and overwhelming time in one's life."

Another student describes her process of choosing Notre Dame for its spirituality after visiting a number of other prestigious schools: "In the process of visiting these colleges, I realized just how important my spirituality would be during my college years. I knew that no matter what school I would choose, I would make spiritual growth a priority. Soon after the East Coast swing, I visited Notre Dame with my mother, and from that first day of the visit, I knew that I would be most happy at Notre Dame."

Students at Notre Dame are service-oriented and want to get involved. Says one junior, "A criticism of Notre Dame could easily be that it is isolated and out of touch with the world outside the Dome. This problem, however, continually challenges Domers to reach beyond the isolation of our gilded existence. Having a Center for Social Concerns on campus and especially programs like Summer Service Projects instills in the students the importance of putting what they're learning in the context of the real world of people and places."

And there is clearly a connection between this service and community focus on the university's religious character: "Three things immediately come to mind as being among the most important things about my ND experience. Dorm life is unique and really shapes the character of the campus. The dorm plays a huge role in the day-to-day experience here and has afforded me wonderful opportunities to develop friendships. An especially important part of that is having mass in the dorms. The Catholicity of Notre Dame, although never imposed, is omnipresent. It also takes a multitude of forms. This unique brand of spirituality is also among the most important things at ND to me. It is wonderful to be in the shadow of the grandeur of the Basilica, to make late-night runs to the Grotto after an SYR dressed in a formal and Nikes, or to go to ten-thirty P.M. mass in the dorm in pajamas and be a eucharistic minister in socks! A part of that sense of Catholicism also takes on a different spin in the presence of the Center for Social

Concerns on campus. The sheer physical presence of the building serves as a constant reminder of the world beyond the Dome. The Summer Service Project has challenged me to integrate the classroom with the outside world. It is all at once emotionally draining and thought-provoking, frustrating, and inspiring. Like much of the Notre Dame experience, it guarantees to be an adventure that will both challenge and enliven."

Another junior says, "PLS is the greatest program ever conceived, also Emmaus and the NDE retreat are essential to the Notre Dame experience. NDE retreat was the best weekend of my life, and everyone who goes on it says the same. It was so special." A sophomore says: "Notre Dame does a remarkable job of encouraging people to become involved with social work during their undergrad years. There are thousands of opportunities for volunteer work (large, medium, and small time commitments available) and thousands of students get involved. They do summer service projects (eight weeks of service in hundreds of locations around the United States as well as some abroad, with a $1,700 tuition stipend and three theology credits), do service trips over fall or spring break, or do random projects during the school year on their own time. Several hundred students dedicate their lives to a year (and sometimes two or three years) of service immediately after graduation. This is a remarkable, admirable quality of Notre Dame students. Graduates aren't simply greedy for great-paying jobs. They are concerned with creating a beautiful, compassionate world, and they want to be active in making positive changes with their skills. In fact, the valedictorian this year turned down a Fulbright

Scholarship in order to do a year of service. Wow—what courage and conviction this woman has!"

Students who do best at Notre Dame balance academic, social, and extracurricular life. "It can be overwhelming to be suddenly surrounded by students from around the country who were all the best and brightest in high school," says a junior. "The key to success is striking a balance. Most people at Notre Dame adopt a work-hard/play-hard philosophy. They are motivated and zealous about everything they do and discover their niche within the community. Everyone at Notre Dame is motivated and I think we all help each other to be the best that we can be. When you are surrounded by people who share your drive and enthusiasm, it makes things easier and more enjoyable. There really isn't the competitive, cutthroat atmosphere that so often hovers over people who want to succeed, and instead Domers operate as a family and community. People genuinely care about each other and all struggle to find a healthy balance."

Another student praises "people who are motivated, creative, and who aren't afraid to make a few mistakes (or lose a few hours of sleep!) in the process of learning and experiencing will be most successful. People who come with an open mind, a positive attitude, and an excitement to get involved with all sorts of things are going to have a fabulous time at Notre Dame. The key is to *get involved*. Without getting involved with something that you love, you could spend your entire time binge drinking or playing video games in a dingy dorm room. There is so much going on! Look at posters! Chill with friends!

Join a club! Volunteer at a homeless shelter! Go listen to guest speakers! See a play! The list is endless. For people who are willing to experience and challenge themselves, Notre Dame is going to shape them into an incredible individual with a beautiful attitude toward life, a great view of the world, and the courage to do something great with their own talents."

Students disagree over the diversity, or lack thereof, on campus. Says a junior, "Notre Dame is easily (and erroneously) criticized for a lack of diversity. There are a multitude of clubs and programs that explore and celebrate various cultures, ethnicities, and backgrounds. Also, each dorm has a multicultural commissioner who is responsible for organizing events within the dorm and increasing awareness and education about different issues." But another notes that "the school has failed in its acceptance of homosexuals, just research and you'll find out. Not enough financial aid to *really* make the school diverse. Most of the cultural differences are in the athletes." A sophomore follows up on this, noting that "this year the big debate was over whether to include a clause about homosexuality in our 'nondiscrimination' policy. We are one of the last universities not to include this clause, and the faculty, graduate student, and undergraduate student committees all voted in favor of its inclusion. Despite the opinion of the majority of students and faculty, the administration (trustees) still voted not to include it." This student continues by noting that "Notre Dame is blatantly a white, Catholic, conservative, upper-class school. Diversity is definitely not its strong point, and while many might find that grounds for criticism, I accept and respect it as the characteristic of this university. In all statistics and surveys, I am considered a minority, and especially at Notre Dame, from superficial appearances, I am one. However, I have never felt part of the minority population. I've always felt included in the entire student body. Notre Dame is not about diversity, it is about unity. As a very traditional and Catholic school, Notre Dame attracts a certain type of person. We all come from basically the same backgrounds, have the same values, and believe basically the same things—even all the so-called 'minority' students. What makes this university unique is that it emphasizes unity instead of diversity; Notre Dame is about promoting togetherness based on the same values, beliefs, and backgrounds."

Other critiques of the university center on financial aid, the role of athletics, and the pros and cons of dormitory parietal rules. "There is so much money available to the school, and yet so much of it goes to things other than financial aid for the students who desperately need it. I think it's fine that Notre Dame has an incredible athletic program, but sometimes it becomes too much of a priority. There are so many athletes who get full rides to Notre Dame, but there are zero scholarships given for academics. Not only do athletes get full rides, they get free tutors for all four years, free books, free everything when they go play in different places. I understand that athletes work very hard to maintain good academics, but there are certainly non-athletes who are working just as hard and getting absolutely no rewards for their work."

Incoming students should be aware of the strict dormitory rules at the university, and how they impact social life: "Gender relations at Notre Dame are very strained. I think all of this stems from the parietal system, which is basically a sexual curfew for all students. Due to the single-sex dorms combined with some very conservative, Catholic sexual guidelines (a.k.a.: no sex before marriage) opposite sexes must be out of the other sex's dorms by midnight on Monday through Thursday, two A.M. on Friday and Saturday, twelve-thirty on Sunday. It's a great way to get parties settled down on the weekends, but it's ridiculous for all other purposes. If people are going to have sex, they will have sex regardless of the Catholic traditions of the school or any sort of parietal system. In essence, the parietal system seems to prevent the natural friendships that develop between men and women when they live in the same dorms. Guys can't just drop by my room without calling up. And even if I'm just casually visiting a friend or actually studying (which happens a lot in real life at ND) the guy has to leave my room . . . Parietals make the opposite sex a little bit too mysterious to allow for a healthy amount of dating and casual friendships that develop outside classes, parties, and activities."

Much of the bonding and social interaction at Notre Dame occurs in the dormitories, facilitating a friendly and familiar atmosphere on campus: "The housing system is wonderful at Notre Dame. Students are randomly placed in a dorm in which they will stay for all their years on campus. Almost all students stay in the same dorm for all four years, although half the senior class moves off campus. This facilitates further community growth, while preventing any sort of exclusive attitudes among students in the way a frat system might. Every student automatically belongs to a dorm which has a specific personality and set of traditions, but there is no rushing or stressing involved. Everyone belongs and everyone can be involved."

Students leave Notre Dame feeling loyal members of a family. "Notre Dame is an institution that is very much alive—it lives through the people who love it and have made it their own. Everyone who comes here invests a piece of themselves into it. Each new class reaffirms that character—it is not a matter of maintaining some archaic rite or ancient lore, but of evolving and growing, grounded firmly in history and tradition. This is what makes Notre Dame great. Notre Dame becomes a way of life and charges its graduates to carry within them the spirit of the university and an undying love for the school and family." A sophomore says, "There is something magical about Notre Dame. You can sense it in the students, in the tourists, and in the air. A humble pride (or in the case of football, a blatantly proud pride) is a part of the Notre Dame experience. I felt it and was attracted to it. . . . I chose Notre Dame because I wanted to spend four years of my life growing, experiencing, and learning in a place that creates compassionate, successful, educated people. I wanted to go to a school I could fall in love with both inside and out."

The "family" feeling at Notre Dame is prominent: "There is a true sense of community among the campus as a whole. And from my experiences this year, I saw the family respond to so many needs with endless amounts of love. Just as any family drops everything to help someone in a desperate situation, I was astounded to see the way students responded to several tragedies during the year—one boy with sudden brain surgery, a girl with cancer, deaths among close relatives, etc. Absolute strangers made banners, wrote cards, came to prayer services—people actually cared about fellow students." A sophomore raves that "Professors are amazing, honest, and excited to help you grow. The religious figures are also very supportive and crucial in making Notre Dame a thriving environment. The most important asset to student success, however, is the student body itself. Students create the unparalleled charisma and passion that is present at Notre Dame. Please do not come to Notre Dame just because you think it will look great to some company in the future. Come because you are drawn to the passion of the student body, to the bold pride in the importance of spirituality and compassion. Choose your school because it will be perfect for you, not because it will look perfect to other people. Don't choose your school because you want to impress relatives at your graduation party or students in your high school. It won't make you happy."

What Happens After College

Service-oriented positions and graduate education are the main choices for Notre Dame graduates. Some 10 percent leave Notre Dame and engage in some form of community service, such as the Peace Corps or the Jesuit Volunteer Corps. About 6 percent join the military as commissioned officers. A large percentage (40 percent) immediately enter law, medical, dental, and other graduate schools. Notre Dame graduates also regularly earn prestigious scholarships and fellowships, such as the Rhodes, Marshall, Truman, Stevenson, and Mellon awards.

OBERLIN COLLEGE

Carnegie Building
101 North Professor Street
Oberlin, OH 44074–1075

(800) 622-OBIE/(440) 775-8411

www.oberlin.edu

Number of undergraduates: 2,900
Total number of students: 2,900
Ratio of male to female: 42/58
Tuition and fees for 1999–2000: $24,264
% of students receiving financial aid: 58
% of students graduating within five years: 80
Endowment: $435,000,000
Endowment per student: $150,000

Overall Features

The Protestant missionaries who established Oberlin in 1833 for the express purpose of training teachers and future missionaries who would go out into the world and do good works would be pleased with the present-day tone of the college. Oberlin was the first male college to admit women and among the first to recruit African Americans to the campus. Its long history of social action and intellectual progressive thinking has continued to define the academic and social tenor of this distinguished institution. A considerable portion of its resources are allocated to financial aid, in keeping with its commitment to educating students of all backgrounds and persuasions. Over half of the undergraduates receive aid and a full fifth of the student body are minorities. Somehow Oberlin combines a liberal, open-thinking environment with a highly academic program while maintaining a spirit of goodwill and cooperation among the students and faculty. The campus is always buzzing with passionate debate and action over issues of concern. This is clearly a campus with a heightened concern for the right to hold unpopular opinions and foster multiculturalism in all its modern forms. It informs the social life, the classroom content, and teaching. Special interest groups along gender, racial, and political lines abound. There are a good number of affinity houses for students who wish to identify themselves with a special category ranging from an intellectual interest to race and ethnicity to sexual preference.

The glue that holds this collection of diverse individuals together is the ethos of the community. There is respect for both the assemblage of all the different people and their beliefs and for the exceptional education available. Oberlin attracts outstanding faculty who delight in teaching and mentoring talented, independent, nonconformist learners, who are in the majority. Great weight is given to the quality of teaching and the personal direction of students as they explore their particular interests. There is a good deal of work assigned by most professors, which students are expected to keep up with while they go about their many outside activities. Oberlin is unique among the top liberal arts colleges for the presence of an outstanding music conservatory on campus. This enhances the cultural life of the college community, which is significant since Oberlin is located in a small town in Ohio.

What the College Stands For

"Think one person can change the world?" asks Oberlin. "So do we." Oberlin stands proud about its egalitarian ethos and historical legacy of progressiveness in American education. The college's statement of goals and objectives, adopted in 1977, outlines its goals for its students: "To equip them with skills of creative thought, techniques, and critical analysis which will enable them to use knowledge effectively; to acquaint them with the growing scope and substance of human thought; to provide for their intensive training in the discipline of a chosen area of knowledge; to ready them for advanced study and work beyond the college years; to foster their understanding of the creative process and to develop their appreciation of creative, original work; to expand their social awareness, social responsibility, and capacity for moral judgment so as to prepare them for intelligent and useful response to the present and future

demands of society; to facilitate their social and emotional development; to encourage their physical and mental well-being; to cultivate in them the aspiration for continued intellectual growth throughout their lives."

Oberlin seeks diversity in its student body and its intellectual approach to learning and sees mutual benefit in the interaction between its College of Arts and Sciences and its Conservatory of Music. As one administrator put it, "Different people would probably have different answers" to the question of what Oberlin's educational goals are. But he suggested: "To help people learn how to learn; to create talented writers, scientists, historians, etc.; to develop students' passion for learning; to educate people to make a difference, make their mark; to challenge students to the fullest."

Curriculum, Academic Life, and Unique Programs of Study

The curriculum reflects the progressive philosophy of the college in a number of ways. Students are required to complete three courses in the four traditional disciplines, but have great flexibility in choosing from a diverse and generous palette. A course that focuses on multicultural topics and issues is required of all students. A January term also is required of all students, which allows them to complete an independent project under the guidance of a faculty member or to undertake a meaningful internship off-campus. No one would question the presence in most classes of a liberal (some would say political correctness defines the attitude) perspective by which most courses are taught. While many students are keenly interested in the top-flight social sciences, humanities, and classics offerings where they can study and debate the hot issues of the day, many others major in one of the several outstanding science departments. Oberlin has a well-deserved reputation for its biology, chemistry, physics, and new biopsychology and neuroscience programs. Similar to a number of its sister liberal arts colleges, Oberlin prepares a disproportionately large number of students for graduate studies in medicine and science teaching and research. Studying abroad is a popular option for many students, which the college encourages through its own programs or through cooperative arrangements with other major institutions. The music conservatory retains its own faculty and intense curriculum that prepares its highly selective student body for performance or teaching careers. A beautiful art museum on campus offers students the chance to work and study in the fine arts. Students in the college of liberal arts can readily elect courses of interest in the conservatory. A key factor that alleviates the demanding academic pace is a policy that erases all grades of C- or lower from the official transcript and allows students to take as many of their courses on a pass/fail option as they wish. This is a popular policy since a majority of Oberlin seniors continue their education in a wide variety of graduate programs.

A strong community service program, an honor code, the student research opportunities and winter term, and the double degree program in liberal arts and the music conservatory "have been important," says an administrator, "in terms of both academic challenge and stimulation and also Oberlin's goal of helping to create leaders and good citizens." The college fosters diversity and divergent views on campus, providing a great deal of support for individual students. "Oberlin prides itself on its diversity. We were the first coed school and the first college to admit students regardless of race. Our history has helped us to maintain diversity, but it is also a major priority for us in admissions and administration. Oberlin continues to have a strong multicultural resources center and a huge variety of student organizations that promote all kinds of diversity."

Major Admissions Criteria

The admissions office at Oberlin looks for "a strong academic background, and also a willingness to be challenged, a love of learning, and an open mind. The people who do best are students who challenge themselves, are willing to learn new things, and work cooperatively rather than competitively." The three most important admissions criteria, according to one member of the admissions office, are "high school grades, choice of classes, and overall intellectual nature and 'fit' for Oberlin."

The Ideal Student

This is an easy call for the high school senior in search of the right college match. Oberlin is not the most appropriate place for the conservative personality or the very traditional learner, nor is it likely to appeal to the committed athlete who wants a competitive intercollegiate sports program. Those who are attracted to the concept of a highly diverse community of individuals with as many points of view as there are students will find Oberlin an exciting opportunity. Respect for the serious intellectual education to be obtained from the excellent faculty will result in a most fulfilling four years that will leave a lifetime imprint on one's actions and view of the world. As one administrator noted, successful students at Oberlin have a "love of learning, passion for ideas, a willingness to be challenged, and an open mind (both socially and academically)." Environmentally, students must be comfortable with life in a small, pretty, self-contained town, even though the college is only thirty minutes from the growing metropolis of Cleveland.

Student Perspectives on Their Experience

Students choose Oberlin because of its academic reputation, its "legacy of social engagement," contact with faculty, music program, and location. As one junior notes, "Oberlin offers opportunities to work with faculty closely in an academically challenging environment. Also, the presence of the Oberlin College Conservatory promotes frequent, high-level musical performances, as well as opportunities for non-Conservatory students to take music lessons from Conservatory students and faculty." Professors earn much praise for their accomplishments and availability. Says one student, "Many of Oberlin's faculty members are amazingly masterful in their fields. Learning from these professors is a wonderfully enriching experience." A sophomore says, "My decision to attend Ober-

lin was based on my visit there. As soon as I stepped out of our car, I knew that Oberlin College was the place for me. It just felt right. I felt incredibly comfortable and had no doubt I would do well. There was an energy to the campus that was infectious. I was excited to think I could be a part of it. My college search ended that day and I applied early decision."

Those who do best succeed because they are smart, involved, hard-working, and tolerant. "Students who do well at Oberlin are reasonably smart students who are willing to work hard enough to be academically successful. The Oberlin environment prevents no capable students from succeeding. Close contact with faculty, numerous opportunities for undergraduate research, and the absence of a 'cutthroat' competitive environment help students succeed." A sophomore says, "The people who do best at Oberlin are those who take a personal interest in the things that are going on. Whether those be on-campus such as class, student organizations, or committees, or off-campus such as local, state, national, or international events, the students who want to get involved are those who do best." A junior says, "Those who use the four or more years at Oberlin to explore themselves and get to know people, leaving the campus regularly to interact with the broader community, do best at Oberlin College." A member of the class of 2001 identifies "the kind of person who cares about learning and helping other people. Someone who takes action. Someone who knows how to study intensely and work hard for reasons beyond 'making the grade.' Someone who makes things happen rather than lets them happen." This student counsels her peers: "Don't feel bad about yourself because everyone around you seems as talented or more talented than you. Instead, find inspiration through these incredible people. Work hard and find reasons to care about what you're doing." A sophomore says that Oberlin "creates a safe environment where students are able to explore not only themselves but also any potential interests they may have without fear of social reprisal. Individuality is cherished and a person is valued for who he or she is. The school also does an outstanding job at piecing together a group of exceptionally diverse people in its student body."

Diversity is seen as significant and positive at Oberlin. One junior says that "all types of diversity exist. Oberlin is extremely accommodating. For example, there are special dorms, such as 'Third-World House,' 'African Heritage House,' and a collection of language program houses that promote diverse cultures." Nevertheless, some students would like to see even more recruitment of students of color, less self-segregation by students (a common complaint across college and high school campuses), and the development of more curricular diversity and flexibility. But the college "creates an atmosphere where students take initiative and do incredible things, such as winter term projects," and where there are experimental college courses taught by students. One student praises "Oberlin's tolerance and respect toward marginalized people, especially gay and lesbian students. Oberlin has always been ahead of the times by being a place that welcomes and supports people who are discriminated against and persecuted by the majority of American society." A sophomore says, "Oberlin College is the epitome of diversity. If you could make a

concentrate out of Earth, as you would an orange, Oberlin would be it. Every kind of diversity, whether cultural, ethnic, racial, socioeconomic, intellectual, religious, geographic, or experiential, exists in abundance on campus. Because the people (not only students but professors and staff) are so different, everyone is very open-minded and sensitive to the needs of others. What makes Oberlin so special is the people. The students, the professors, the administrators, and the staff are some of the most amazing people I have ever met. We are all vastly different but are brought together in this one place where we are encouraged to grow both as individuals and as a community. I have just started my second year here but I already feel I will never truly be able to repay Oberlin for everything it has helped me become. If you're not afraid to be challenged or shocked, and want to get involved, come join us because Oberlin College is where you belong."

A junior reflects the college's accomplishment of its positive message about changing the world: "The amazing, talented, inquisitive, and socially concerned students who go here make the college special. Students tend to help one another rather than view each other as competition. The Oberlin Student Cooperative Association (with a big emphasis!!) offers the unique opportunity for students to control and operate programs that have dramatic, positive effects on the community. The lessons you learn at Oberlin encompass more than academic courses. You can find hope here, knowing that with dedication and hard work you can make positive change."

What Happens After College

Oberlin students enter myriad careers and fields after graduation. As one administrator puts it, "While about sixty-five to seventy percent of our students eventually go to graduate school, the jobs and fields they pursue run the gamut. They are prepared to do just about anything. The most popular career choices include education and communications." Indeed, higher education (17.5 percent) is the number one occupation of Oberlin graduates (Oberlin has one of the highest numbers of Ph.D.s among the small colleges), while over 12 percent each enter health professions or business and commerce, 9 percent go into public service, 8 percent into music performance, and 7 percent into K–12 education. Oberlin maintains an active alumni network, and over 1,600 internship opportunities through its Business Initiatives Programs, factors which help students secure jobs after graduation.

POMONA COLLEGE

Summer Hall
550 North College Avenue
Claremont, CA 91711-6312

(909) 621-8134

www.pomona.edu

[
Number of undergraduates: 1,500
Total number of students: 1,500
Ratio of male to female: 52/48
Tuition and fees for 1999–2000: $23,170
% of students receiving financial aid: 55
% of students graduating within five years: 93
Endowment: $675,000,000
Endowment per student: $450,000
]

Overall Features

One of the oldest independent colleges in the West, Pomona was modeled after the traditional New England schools like Amherst and Williams. Over the course of its 100 years as the leading liberal arts college west of the Rocky Mountains, it has taken on its own special identity as a reflection of its mission and its location. In addition to stressing the importance of small classes and a commitment from the faculty to teach students the critical skills that distinguish a liberal arts education, Pomona prides itself on the highly selective and unusually large number of racial and ethnic minorities in its student population. Few private institutions can match the 40 percent minority census of Pomona. The largest number of minorities represent the many different Asian heritages. Pomona has a significant endowment that allows the admissions committee to practice need-blind admissions in their decision-making process, a growing rarity among high-priced private colleges today. The result is a student composition of highly talented young men and women of a wide range of economic and racial backgrounds. The student body is a diverse, liberally minded, intelligent cohort of 1,500 undergraduates. Pomona's small population ensures a close community of students and faculty and small classes.

At the same time, the potential for social and intellectual claustrophobia in such a setting is mitigated by the unique consortium that comprises the five Claremont Colleges. As other colleges were created in the Claremont group, they modeled themselves loosely on the Oxford University college system. Each of the five schools has a particular academic focus and pedagogical style. Claremont-McKenna emphasizes the social sciences,

Scripps the humanities and arts, Harvey Mudd the physical sciences and engineering, and Pitzer the behavioral and social sciences. Students have access to all courses and activities sponsored by the other colleges. Thus a small college can expand for the enterprising student to an intellectual and social community of over 5,000. A new recreation facility has become the center of action for students and tends to keep them on campus.

Students and administrators consider the college a vibrant and serious place to develop intellectually and personally. There are only a handful of fraternities on campus and no sororities. They do not have a major impact on the social tone of the campus. The large endowment has created an extraordinary set of majors to choose from, an excellent faculty, and many extracurricular organizations to join. Athletics are given a good deal of support as well: Pomona fields nineteen varsity teams that compete in Division III. Several, such as beach volleyball and inner-tube water polo, reflect the culture and locale of sunny Southern California. Pomona administrators, teachers, and students believe they have the best of all worlds: a small, caring academic and social community with excellent resources for learning and a student body that is representative of the demographics in the western United States; at the same time Pomona has the advantages of a mid-size university because of its membership in the Claremont Group. All of this exists in a locale where the sun always shines. Although Los Angeles is only forty miles away, some students state that they rarely venture into that sprawling urban center because they have what they want on campus or do not own a car, which is the only way one can get around.

What the College Stands For

Pomona maintains among its faculty and staff an intense commitment to undergraduate education in the liberal arts tradition, focusing primarily on quality teaching and interaction between students and faculty. The college sees itself as part of a very workable college consortium that greatly expands academic, social, and extracurricular opportunities for its students. Its location near Los Angeles and its proximity to the Pacific Rim are held up as major strengths and defining aspects of the college. Pomona is committed to diversity on campus and backs that aim up with a lot of aid, financed by its large endowment. About one of every fifteen students comes from a family with income less than $20,000, and about one in five are first-generation college students. A senior administrator expresses his vision of the college in this way: "I would say that Pomona has a number of powerful elements of uniqueness and distinctiveness. Among America's premier liberal arts colleges, we are the only one in the western part of the country geographically proximate to the Pacific world and Latin America and located on the edge of arguably the most dynamic city in the country, Los Angeles. Southern California is one of the richest teaching laboratories in the world, and we use our placement here to great advantage. Pomona is also distinguished from other leading colleges by its role in founding the Claremont Colleges, five colleges and a graduate school within walking distance of each other. This enables our students and faculty to combine the benefits of a small college with those of a larger setting, giving us a critical mass of faculty in selected fields, adding to the diversity of students and cultures, and providing a university-level library far superior to that at any individual liberal arts college I know. Students find that their neighbors in the other Claremont colleges also enhance social life. Finally, Pomona is distinctive because it is one of the very small number of colleges that really is need-blind in admissions. Talk is cheap, but very few other institutions today can honestly say that they accept students entirely on the basis of merit and promise and then provide the financial assistance necessary to permit those students to enroll and to prosper at the college."

Another senior dean says, "Our focus is solely on undergraduates and our primary educational goal is to provide an undergraduate education that is distinctive and better than any other. Within that context, we seek to develop students' critical capacities and to stimulate them to achieve academically at the highest level. In all of our educational endeavors we stress adaptability and flexibility, i.e., knowing how to go about finding solutions rather than solely developing proficiency in given solutions. Our effort is based on four pillars: the major, the senior exercise, the Critical Inquiry (First-Year) Seminar, and the Perception, Analysis, and Communications requirements for graduation. We seek to prepare our students for leadership and for service and to participate in a number of roles in the larger world. We seek to develop an eager embrace of pluralism, both domestically and globally. Importantly, we focus on the development of a global consciousness and an ability to understand and to participate in a variety of cultures and environments."

Says one administrator, "Pomona's educational goals are to promote in our students very high levels of academic achievement, to cultivate in them the capacity to live responsibly and creatively in a diverse society, to prepare them for advanced study, and to equip them for positions of leadership. We pursue these goals not only through formal instruction but also through the character of residential life on the campus."

Curriculum, Academic Life, and Unique Programs of Study

Pomona sets high academic standards through rigorous coursework at the same time that it gives students a good deal of freedom to fulfill the classic disciplinary areas of study. There is a three-semester foreign language requirement that can be waived by demonstration of competency. All first-year students take a critical-inquiry class or an interdisciplinary seminar. All seniors are required to complete a thesis or project related to their major. In addition to the large array of courses and concentrations available, students can create their own major field of study under the tutelage of a professor. About one-fourth of students at Pomona major in an interdisciplinary program. The underlying goal is to ensure completion of the ten essential skills of learning and thinking the faculty believe are necessary for further study and leadership roles in the larger world. Strong departments are English, economics, history, and life sciences. Pomona has a well-deserved reputation for preparing students for medical school successfully due to the excellent teaching, research opportunities, and small classes. A science major is not likely to find himself learning chemistry or biology in a lecture class of 200 to 300 students. Says one dean, "Pedagogically, we are more and more emphasizing discovery-based or research-based learning as well as collaborative learning." Members of a typical senior class will move on to many of the nation's best graduate schools of medicine, business, law, and the social sciences. The majority of students have a purposeful attitude regarding their studies; at the same time they regard themselves as somewhat laid back in style and in concern for grades and competition.

One administrator raves about the changes and developments on the campus in recent years: "Our facilities are better than ever: a wonderful new campus center recently opened; the renewal of science facilities is under way with the construction of the Andrew Building for Physics, Mathematics, and Computer Science, which will be followed by complete renovation of Seaver North chemistry building; the new Hahn Building and the award-winning renovation of the historic Carnegie Building put all our social science teaching in a completely modern, well-equipped space; the construction of Seaver Theater, which has been described as the best undergraduate teaching theater in the western United States; and the complete renewal of athletic facilities occurred within the last decade. On a different front, our student body is more national, more ethnically diverse, and more highly qualified academically than it was only a few years ago. Other changes of note: Our teaching is more research-based and discovery-based than lecture-oriented; interdisciplinary study has grown to involve roughly

twenty-five percent of majors; our faculty, as well as our student body, is more diverse than in the past; and substantial growth and improvement in the other Claremont Colleges enable them to serve our students better through cross registration." In fact, students at Pomona may choose, at no additional cost, from 2,500 classes taught at any one of the Claremont College campuses annually.

Students at Pomona benefit from numerous academic and residential life programs that connect them to the campus, to faculty, and to internships. These include college facilitation of faculty living on or near campus, training for faculty in advising, matching of appropriate faculty advisors with students, a student sponsor program matching sophomores with first-year students, numerous clubs and resource centers, and the Summer Undergraduate Research Program, which is funded by the college.

Major Admissions Criteria

Pomona's admissions process has become increasingly selective, with recent statistics indicating that 83 percent of the freshman class were in the top 10 percent of their high school class and had a median SAT score of 1,420. Pomona seeks diversity and does not consider financial need in its admissions decisions. The college identifies its most important admissions criteria as follows: "The strength of the high school program, the record of academic aptitude and achievement, and school recommendations are very important. Special talents or experiences in science, music, drama, art, journalism, athletics, community service, and other fields, as well as evidence of originality, energy, motivation, and leadership potential are also considered in the admissions process."

The Ideal Student

The challenges one will encounter at Pomona belie the beautiful setting and pace of Southern California. Pomona selects intelligent academic performers who want a first-rate education and who will take a good deal of initiative in planning their studies and overall experience. They also look for the more mature candidate who will embrace the great mix of student types and the open-minded attitude that prevails. This is not an environment for the social phobic who will feel comfortable only in a conservative residential community of similar types of students. The individual who looks at all the opportunities available to him will find Pomona an ideal place to spend four years. There is a plethora of social, political, and athletic activities in which to find one's niche while gaining a first-rate education and a jump-start to graduate school.

The college is challenging but supportive, as one administrator describes: "Students succeed at Pomona because they have it in them to do so: We choose them with this in mind. But other things help, as well. Pomona's faculty (for all their scholarly accomplishments) are first and foremost teachers. Students benefit from taking advantage of this and making the most of the ease with which faculty can be engaged. Students also benefit from our financial-aid policies, which are need-based, and our commitment to meeting their full documented need. This takes a source of anxiety off the table. Student culture helps sustain our students, because while very achievement-oriented it is *not* aggressively competitive: People help and support each other, and there is a lot of student collaboration with faculty and with each other. Finally, I am impressed by the system of safety nets we have in place for our students, overlapping patterns of advisers, mentors, sponsors, clinics, and the like, some of which are functional, some of which are identity-based, and some of which are residential. A student who is in trouble has a very high probability of getting picked up before things get out of hand."

Student Perspectives on Their Experience

Students choose Pomona because of its small size, location, connections to the Claremont Colleges, faculty interaction, diversity, and academic reputation. Says one senior, "I chose Pomona in part because it is a small college. Going to Pomona I feel as though I belong to a family or community. The administration, faculty, and staff care about me. I don't feel this is true about every college. I enjoy having small classes and access to professors. I also appreciate the fact that all classes are taught by professors and not graduate students. Because there are no graduate students here there is no need for competition between them and undergraduates for resources, professors' time and attention, and research opportunities. Although Pomona is small it still has a lot to offer by way of majors, clubs, and other opportunities. We are close enough to Los Angeles that students can take advantage of all a large, diverse city has to offer. I also wanted the opportunity to compete in athletics at a competitive level in college without sacrificing academics. Also, being in Southern California is great for training."

Another senior says, "I chose Pomona for several reasons. First, I was looking for a small, academically rigorous, liberal arts school. Pomona offered this and more. Because it is part of the Claremont Colleges consortium, it offers all the benefits of a small school with the resources of a larger university. The Claremont Colleges, for example, pool their resources for the main library, which contains about two million volumes. Additionally, I can take advantage of the guest speakers and social functions that occur on other campuses while still taking small classes with a lot of student/professor interaction. The second main reason I chose Pomona over other highly selective liberal arts colleges was because of its noncompetitive academic environment. Pomona students do not compete against each other but rather work collaboratively and collectively both in and out of the classroom. Third, extraordinary diversity among the students and faculty drew me here. Not only does Pomona have the highest percentage of minority students of the colleges I considered, it is committed to diversity of all kinds."

Pomona challenges and encourages its students, without overwhelming them. "Pomona is best at allowing and facilitating the growth of individuals in many different ways, academically, athletically, and ideologically," says a senior. "This was also important to me when I was picking a college. I wanted a place where I could feel comfortable."

Those who do best at Pomona are "students who are willing to work hard, ask for help, and are open to new ideas. After students have

graduated from Pomona they are not only expected to have retained an enormous amount of information about their major but also to understand the way it is connected to all other disciplines. Nothing exists in a bubble." A senior says, "The students who seem to do best at Pomona are those who were well-rounded in high school and want to continue exploring a variety of interests in college. Balance, therefore, is a key factor for those who are happiest at Pomona. Open-mindedness is another characteristic of highly successful Pomona students. No matter where a student is coming from, he or she will encounter a variety of different people with views and experiences that are unfamiliar. A willingness to learn from others' unique experiences will also lead to much success at Pomona. Finally, motivated students are bound to excel here. Pomona is very tough academically, and a lot of motivation is needed to get the most from your Pomona experience. Since these are some of the qualities that the admission staff is looking for, it is no wonder that Pomona students and alumni are so successful."

In addition to their peers and professors, students find support through a number of campus resources and programs. "When I arrived at Pomona," says a senior, "I had already been assigned to a Sponsor Group. This is a group of fifteen or twenty freshmen and two sophomores. It becomes your base. They match people up according to interests, so many of my closest friends were in my Sponsor Group. It's like a family unit. The sponsors are your mentors and can help you with basically anything. I was also assigned a mentor from the Office of Black Student Affairs (OBSA). She was basically another person that

I could go to, and she also introduced me to a lot of people. Both of these programs helped to ease my transition. I was also assigned an academic advisor who helped me pick out classes and eventually choose a major."

Finally, a word of advice about name recognition, which could be applied to the Hidden Ivies in general: "Prospective students considering Pomona often struggle with what seems to be Pomona's lack of a reputation. Often I hear prospective students say, 'No one at my high school has ever heard of Pomona.' True as this may be, Pomona's academic excellence speaks for itself. The overwhelming amount of personal and career success that Pomona alumni achieve is proof that the college has an outstanding reputation. Keep in mind that graduate schools and top employers *do* know about Pomona!"

What Happens After College

Pomona's graduation rates are among the highest nationally, and the college's attrition rates among the lowest. Graduates go on to succeed in many different professional and graduate programs, including medical schools, law schools, and M.B.A. and Ph.D. programs. While a higher percentage eventually pursues graduate degrees, 27 percent of the class of 1998, for example, went directly to graduate programs. Of these, 19 percent were medical schools, 17 percent law schools, and 58 percent other graduate schools. Of the 63 percent of the class of 1998 that went into employment, 31 percent were in business, 18 percent in education, 15 percent in arts and communication, 12 percent in science, and 12 percent in non-profits.

REED COLLEGE

3203 SE, Woodstock Boulevard
Portland, OR 97202

(800) 547-4750

www.reed.edu

Number of undergraduates: 1,300

Total number of students: 1,300

Ratio of male to female: 47/53

Tuition and fees for 1999–2000: $23,050

% of students receiving financial aid: 50

% of students graduating within five years: 70

Endowment: $247,560,000

Endowment per student: $190,430

Overall Features

Whenever reference is made to the progressive model of higher education, Reed is always one of the top colleges mentioned. Reed was founded in 1908 to offer a more flexible, individualized approach to a rigorous liberal arts education. This is an important point to make because many people assume that the academic structure and student culture that personify this liberal institution of today grew out of the student revolts and experimentation of the late 1960s. In fact, Reed has remained true to its original mission to provide intelligent, intellectually passionate young men and women with a first-rate education in an atmosphere of free inquiry and reflection. Any changes in programs or requirements over the decades have had to meet the test of this philosophy. Too casual an interpretation of the quintessential Reed student and the culture of the community can lead to some wrong conclusions.

While students take pleasure in presenting themselves as free spirits, nonconformists, and intellectuals in search of the eternal truths, they are very serious students who relish meeting the rigorous demands of the curriculum and the high standards set by their professors. Classes are quite small and the opportunities for independent research and writing are considerable. A premium is placed on the quality of teaching and advising since intellectual dialogue and study is the key point of the Reed experience. Teachers are evaluated by their professional peers and students on a regular basis in determining tenure. One chooses to teach at Reed because of a passion for one's discipline and the desire to transfer this to students rather than to engage in advanced research as one's primary activity. Typical Reed students gain such a mastery of the subjects they choose to study by virtue of the intellectual skills required and the opportunities to work hand in glove with their teachers that the track record for entrance to the top graduate schools is extraordinary. For example, Reed ranks second among all liberal arts colleges in the production of Ph.D.'s and third in the percentage of graduates in science and engineering who earn Ph.D.'s. Only Cal Tech has a higher ratio of graduates who earn doctorates in all fields. Thirty Reed alumni have won prestigious Rhodes Scholarships for advanced studies at Oxford University. A significant number of candidates qualify for medical school entrance. These are impressive results, indeed, for a college of only 1,300 undergraduates. There are no fraternities or sororities on campus, and the athletic programs are noncompetitive intramural programs. A great many students will study abroad for one or two semesters irrespective of their field of concentration. Reed operates on an honor code system that gives students the freedom to monitor their own behavior academically and socially. A student judicial review board deals with infractions and makes recommendations to the administration. Only 60 percent of students live on campus due to housing limitations and the availability of alternatives in Portland, a small, diverse, and interesting West Coast city. All first-year students are guaranteed housing, however. The fact that one-half of all students receive financial assistance reflects the diverse mix of individuals who make up the Reed community.

What the College Stands For

"There is a dissonance," a senior dean told us, "between the curriculum at Reed, which is formal, and the social environment, which is not." This is the central identity of Reed College, an institution with a demanding set of core courses and a thesis requirement. These aspects of the college create an "intellectual community" where all students are completing the same readings at the same time as all other students. Students are encouraged to work together and be vocal in classes taught in a "conference method." The college seeks and nurtures "intellectual" students in a very liberal and progressive social atmosphere. Reed sees itself as a "unique, intellectual place, with a long-standing and serious commitment to the liberal arts," with more kinship academically to such institutions as Columbia University, the University of Chicago, and St. John's than to Bard, Bennington, or Hampshire. As its catalogue says, "The goal of Reed College is to provide an education in the liberal arts and sciences with emphasis on the highest intellectual and scholarly standards. The Reed education pays particular attention to a balance between a broad study in the various areas of human knowledge and a close, in-depth study in a recognized academic discipline."

"From the beginning," say college officials, "Reed has sought to provide its students with the most serious and rigorous undergraduate education in the liberal arts and sciences. The college was explicitly founded in reaction to the then prevailing model of East Coast, Ivy League education. At Reed, there would be no varsity athletics, no fraternities, no exclusive social clubs: The college would be nonsectarian, coeducational, and egalitarian in spirit. The focus would be strongly academic and intellectual.

The idea was to create an intense intellectual community, devoted to the life of the mind, in which students would be expected to attain the highest degree of undergraduate scholarship." That philosophy has remained in the forefront of Reed's identity and culture today. And in the future? "Reed will continue to remain uncompromisingly committed to the highest undergraduate intellectual standards in American higher education. To that end, the college is striving to do even better what it has always done exceedingly well: to support an environment where faculty and students engage in an academic program of rigorous collaborative learning. To enrich the learning environment, for example, the college has made a commitment to further reduce the student/faculty ratio to ten to one by adding several new tenure-track faculty positions without increasing the size of the student body. Reed has also undertaken an aggressive $100 million fund-raising campaign."

Curriculum, Academic Life, and Unique Programs of Study

Reed fosters student independence of intellectual exploration and opinion, but it does so within a curriculum that ensures exposure to the great ideas and defining events in the history of the civilized world. Students will share in the experience of Humanities 110, a two-semester course based on the great books concept. This entails a heavy reading load of primary sources, which range from the Greek and Latin classics to St. Augustine's *Confessions,* taught by a team of twenty faculty. A majority of students continue on to other humanities topics that cover the Renaissance, modern

European history and culture, and Asian civilization. Beyond this foundation, students are free to choose from a large selection of arts and science courses. There are stages in the four-year education that measure what a student has learned in his field of study. Juniors must pass a qualifying exam before moving on to their culminating work, a senior thesis of considerable depth of thought and content. If the results of the junior qualifying exam are not satisfactory, a student can be required to undertake further study or change his major if the faculty think it advisable. Seniors must also sit for a comprehensive two-hour oral review by their department's faculty that is in essence a defense of their senior thesis. It is no wonder that Reed undergraduates are prepared to tackle the rigors of graduate school studies, which require the same skills and process of learning.

"Over the years," says an administrator, "the college has adhered to these principles with extraordinary fidelity. The fundamentals of the academic program have remained unchanged: a highly structured curriculum, including a required first-year course on classical Greek and imperial Roman civilization; substantial distribution requirements ensuring genuine breadth of study for every student; a required junior qualifying examination, in which each student certifies his or her mastery of the major field of study; a year-long senior thesis, required of every student and culminating in a two-hour oral defense before a committee of the faculty." Philosophy is a popular and highly regarded department. Reed offers an excellent science curriculum for the serious student who is contemplating medical studies, engineering, or research. As an expression of its determination to emphasize self-learning and self-discovery rather than grade accumulation and competition, Reed faculty issue grades in their courses but do not release them to students unless their performance is lower than a C.

While Reed's core curriculum has remained largely unchanged over the years, the college's approach to the core material has evolved. According to one administrator, "For over fifty years our first-year students have been required to study, among many other works, Herotodus's history of the Persian wars. But the ways in which we teach this book have changed dramatically, and now include newfound attention to the issues of material culture, slavery, gender, multiculturalism, and the like. The faculty is professionally active and engaged, and when they teach materials they are naturally disposed to rely on the latest, most advanced scholarly theories and techniques . . . This faculty controls the academic program. In doing so, it has consistently and enthusiastically reaffirmed the traditional pedagogical values of Reed College while at the same time bringing the academic program to the cutting edge of contemporary scholarly and intellectual life."

Another important aspect of Reed's academic life is the college's approach to grading, through which, in the context of a campus-wide honor code, it tries to discourage inter-student competition and encourage meaningful evaluation: "Virtually all work at Reed is graded in the standard way, and, as a statistical fact, there has been essentially no grade inflation at Reed for at least fifty years. But we have an unusual approach to talking about grades. Grades are never affixed to papers or exams,

which are returned with lengthy comments only. Student report cards do not provide specific grades, but only indicate if work is satisfactory or not. Grades are never posted on faculty members' doors. There is no dean's list and no honor roll; students do not graduate cum laude, and there is no such thing as a valedictorian. The idea is simple. We believe that grades are important and that work needs to be rigorously evaluated in the traditional manner. But students should study not primarily for the grade but for the satisfaction of the work itself. This view is best communicated by a set of practices that keep grades more or less private and that do not make a public fetish of them. The results are extraordinary. Grade-mongering is rare at Reed, and students really do commit themselves to intellectual activity for its own sake." Overall, the college maintains "a philosophy of openness and egalitarianism with respect to community life. The campus runs according to an honor principle that governs everyone. There are no closed groups on campus; anyone who wants to join a group or organization can. Students are treated as adults and play a full and equal role in governing their own lives."

Major Admissions Criteria

The college notes, "We realize that there is no perfect combination of test scores and GPA that would predict success at Reed, but certainly a proven track record of academic accomplishment serves students best in the admission process. In reviewing applications, we look for students who are passionate about learning and who have articulated that passion through course selection and achievement. Strong verbal and quantitative skills and demonstrated writing ability are also important considerations. We look for a student body that is diverse in its range of backgrounds, interests, and talents while uniformly committed to a demanding and rigorous undergraduate preparation." The three most important admissions criteria are identified as: "the depth and level of a student's preparedness as demonstrated by the courses he or she has chosen to take and his or her academic success in those courses; a student's love of learning as demonstrated in academic record, faculty and counselor comments, and a student's essay; and an understanding of Reed's unique program and its appropriate fit with a student's particular goals, interests and talents."

Reed's applicant pool is small, self-selecting, and talented. Exhibiting an average GPA in high school of 3.7, 2,018 students applied for the 1999 freshman class. Of these, 1,376 were accepted and 337 matriculated. The college has a fairly diverse student body geographically, with a quarter of the 1999 freshman class coming from the Northeast and significant percentages from California (16 percent), the Mountain/Southwest region (15 percent), and the Midwest (12 percent). However, in terms of international (3 percent) and ethnic/racial minority (12 percent) enrollment, Reed is less diverse. The college says, "Diversity at Reed is considered a multifaceted attribute and is thus not limited to discussions of race. That said, self-identified students of color account for 11 percent of the total student population, and the college actively seeks to increase this number through recruitment efforts by the admission staff, outreach programs involving current students and alumni, and increased campus activities that address multiculturalism. Other 'hot' areas of discussion include gender, class, and religious diversity."

The Ideal Student

Be careful in judging Reed from any cursory description of its approach and style of education or from a campus visit. What can be perceived as a laid-back or countercultural gathering of men and women will not get to the heart of this small community. Reed is an intellectual turn-on for the bright and independent thinker who relishes the exercise of reading the great writers and thinkers or the opportunity to conduct scientific research with dedicated professors. A strong work ethic also defines the successful Reed student. There are a number of academic requirements and hurdles that must be mastered in order to graduate. This is not the right community for a conservative person, socially or intellectually. The faculty pushes its students to think for themselves and to challenge existing concepts and traditions. There is little interest in social conformity, partying, or traditional athletics. The potential candidate for Reed is a lover of books and ideas who is comfortable in his or her washed-out jeans.

College officials say that successful students at Reed "love to ask questions; have strong personal organizational skills; are self-motivated; have a study style well adapted to a rigorous academic pace; enjoy a conference-style learning environment; have a willingness to consult or ask for help when needed; are willing to engage in dialogue with faculty and others; are sustained by an intense curiosity about learning for learning's sake; are able to say no, set limits, stay on track; and have a sense of humor and balance in what can be a stressful culture." Another notes Reed's combination of social and academic environments: "What is most unique about Reed in today's made-to-order world is its unfaltering commitment to tradition. A quick glance at Reed's traditional curriculum proves that the college maintains the classical academic rigor on which it was founded, but one look at the independent-minded student body confirms that Reed is the perfect school for the quirky intellectual. Reed attracts students with a high degree of self-discipline and a genuine enthusiasm for academic work and intellectual challenge, but beyond that there really is no typical Reedie—the community proves to be a place where geeks, glamour queens, and geniuses coexist in academic bliss."

Student Perspectives on Their Experience

Students choose Reed because of its culture, intellectual stance, academic program, location, and faculty. As one graduate says, "It's really the profs that make it so wonderful. They are absolutely accessible, to the point of having students to their houses for dinner *and* accepting invitations from students for dinner at the students' houses. They are brilliant, open, and ready to talk about academics, school, life, or the weather. I think it helps that the profs are here to teach, and that there is no 'publish-or-perish' mentality. They make it great." Adds a senior about her visit to Reed as a high school student, "The campus was beautiful, the staff and students were remarkably friendly and open, and I sat in on

an English conference led by Professor Pancho Savory. He walked in, sat down, read a passage from Ralph Ellison's *Invisible Man,* put the book down, looked around the room, and waited. Eventually, a student made a comment and students proceeded to lead an amazing discussion for an hour and a half. It was pretty exciting."

What kind of students do well at Reed? "Generally speaking, those who do best are self-motivated and have a profound interest in learning," says a senior. "Grades are actually somewhat of a taboo subject among the student body—we just don't talk about them. The pressure to succeed, if there is any, becomes internalized because there is no external competition (with other students). The only benchmark of performance that one has is one's previous performance, and so we are challenged to excel." Another student identifies "people who are open-minded and who want to take responsibility for their own education. Those annoying kids who always answered the questions and/or questioned the teacher." Students succeed because of "flexibility and a willingness to put their nose to the grindstone and *work*. There's no way around it. The only way to succeed in academia is to have a modicum of intelligence and a large will to work; Reed is a true ivory tower of academia."

Reed has a reputation for having students who are lone intellectuals, and some students, including this junior, clearly reflect that opinion of the resulting campus life: "For one thing, Reed's social scene is, on the whole, pretty weak. Reed students come here because they're individualists and nonconformists, and thus big social events tend to be pretty dysfunctional. Nobody here has fun in quite the same way as everyone else. Also there's sort of a mystique here of the angst-ridded, homework-laden, confused young adult. People pretend to be much more bitchy than they actually are. This translates into a campus full of students who for the most part don't look anyone else in the eye and walk with their heads down, even when they pass by people they know from their classes. On the whole, Reed students aren't friendly, or at least they cultivate that appearance. They tend to have a few good friends and remain pretty neutral to everyone else. It's good that the campus isn't full of cliques and so forth, but it would be nice if people smiled more often around here."

A senior says, "Life at Reed can be very isolating at times. We've never quite gotten a handle on how to balance our extremely heavy academic load with other healthy pursuits. The college tries to help students in this respect, but the social climate and the average mentality of Reedies makes this nearly impossible." Students take solace not only in their individual academic pursuits, however, but also in the many available social options available in Portland. Other areas of the college that students would like to see improved include more diverse curricular offerings taught by more faculty and greater student diversity. "That said, however, Reed does have a lot of diversity. Not in skin color or socioeconomic diversity, to be sure, but in a diversity of political opinions, hair colors, tastes in music, sexual orientations, and geographic origins."

A particularly valued annual Reed event is Paideia. "This has to be the coolest student-run event *anywhere*. 'Paideia' is ancient Greek for 'learning' or 'transmission of culture.' The event itself is held during the ten days before the beginning of spring semester, so it's usually the last week in January. During Paideia the student body springs for classes to be held on campus. Usually there are between two and three hundred, and the budget can run to $10,000. Most are one-shots, although some are three—or four—session deals. The classes are about anything and everything. There are lots of cooking classes, there are also classes on Plato. And jazz. And furniture-making. And Tuuvan throat singing. Newspaper hat folding (true class—taught by an older student's six-year-old son). There have also been acrobatics and clowning classes. Essentially, anything goes. They are taught by students, staff, faculty, alums, and random Portlanders. It must be noted that our college's 'most distinguished' alum is Dr. Demento, and he puts on video versions of his famous radio show for the students every year. Those usually fill the hall to capacity, no matter what the capacity actually is. Paideia is just a whole lot of rip-roaring wholesome fun. It's a nice way to learn all kinds of cool stuff and to ease back into the swing of the school year while hanging out with friends."

The honor code is recognized as an important foundation for campus life: "The gist of the honor principle is that Reed students (and faculty, administration, etc.) are expected to be honorable without being forced to be honorable. If you live in a nonsmoking dorm, you don't smoke in there. You can take your final exam back to your dorm room to do it and you are expected not to peek at your books for help, and as far as I know, people are honest about this. They behave this way because they are supposed to, not because there is a formal system of rules and punishments that makes them do it. When someone on campus does something dangerous, stupid, mean, or juvenile, it's the students that usually mete out the punishment, not the administration. As a result, most students here have a lot of respect for the honor principle and get pissed when someone stupid violates it. This way, people learn to be responsible, rather than chafing and rebelling against school policy."

A student further explains the impact of the grading policy on academic life at Reed: "We are given grades but we do not see them. Our professors record grades in their books but every paper/test we get back is heavily accented with comments—which can often be quite copious. What this policy accomplishes is, in my mind, three-fold. One: It removes the competitive pressure. It's hard to compare comments—'Oh yeah? I got three comments.' 'Well, I only had one but it said "good argument."' These things are hard to judge. Two: The benchmark of excellence then becomes yourself. The only way I can judge my performance is by looking at my previous performance, any personal goals I set for myself, and more important how I feel about the work I've turned in. Three: Because grades are actually recorded, we have a transcript to show for our efforts that is readily understandable by outside agencies, such as grad schools. On this point I must add that the college recently completed a study in which 'no grade inflation' was found in the past fifty years. So an A at Reed actually means something. If a student receives a C- or below in a class at midterm grad evals, they get what we jokingly refer to as a 'pink slip' in their box, notifying

them of inadequate performance and encouraging them to discuss any problems they're having with both their professor and their academic adviser. I really like this policy because of the lack of competition and the emphasis on bettering yourself. Ever since my junior year in high school I've been sick-as-a-dog-tired of being used as the class's benchmark, and I am no longer anybody's standard except my own. The policy has a significant effect on the community; instead of talking about our grades, we talk about what we're learning in class. Across disciplines, across major divisions (i.e., science/humanities), our student body discusses what it studies. It creates an unparalleled intellectual atmosphere on campus, one that I relish."

This environment is also facilitated by the conference style of teaching. Says a recent graduate, "People have to be self-motivated and willing to defend their beliefs at Reed. Classes are all conference style, with the professor taking a backseat to student discussion, so you can't just sit and take notes and you *must* do the reading, otherwise everyone would just sit around the table and say nothing. Also, there are no 'sacred beliefs,' meaning nothing is taken for granted. One must support all premises as well as arguments."

Reed challenges its students academically, and in terms of their preconceptions. Says a senior, "The thing that Reed does best, and has done consistently throughout its history, is challenge its students. Or rather, it allows its students to challenge themselves and each other. From day one you're interacting with ideas that are completely different from those that you have known. These ideas come from fellow students, yes, but also from the past three thousand years of the intellectual history of mankind. Through being confronted with difference, the minds of Reedies acquire a tremendous critical skill (finely honed through thousands of hours of in-class and out-of-class practice). This skill is applied to the important texts of our heritage but also to current social issues. By the time Reedies graduate, our minds have gone through a tremendous transformation—and we've developed a set of views and opinions that are our own and have been fully thought out. Somewhere along the way Reed also manages to instill a deep love of learning (if there wasn't one to begin with), which leads almost two-thirds of our alums to grad school."

Another senior says, "Reed will teach you how to think. You learn critical skills—how to write down what you think in anywhere from five to one hundred pages, how to pick out the important parts of what you're reading, and how to discuss questions creatively and intelligently. Reedies can pretty much discuss any subject because we have at least a cocktail party knowledge of everything. Well, not *everything*." And a final piece of advice for prospective students: "Reed is different from every other college and university that's out there. That difference is neither good nor bad. But if Reed is the kind of place a student is looking for, it's paradise. And if you're not into how we do things, it's a very painful experience. I give tours of the campus and one of the things I tell students who seem unsure about coming to Reed is to choose wisely—you won't change Reed, Reed will change you, in more ways than you can imagine. But that's not an experience that everybody wants, needs, or is looking for."

What Happens After College

The majority of Reed graduates (some 70 percent) will eventually earn advanced degrees. The college reports that in a recent alumni survey, "one-fourth of all Reedies reported that they were working in the field of business and industry, with 10 percent identifying themselves as either founder, president, or CEO of their company. This category was second only to education, which captured 30 percent of the alumni and was far ahead of the third-largest categories, a split between arts/communications and health care, both reporting 8 percent of the alumni population." From Pulitzer Prizes to MacArthur "genius" grants, Reed graduates have distinguished themselves in diverse fields, including business and academia. "It is often said," an administrator points out, "that the liberal arts prepares students for anything. Indeed, a Reed education can provide limitless opportunities. A liberal arts curriculum, with its inherent focus on critical analysis and oral and written communication, provides a breadth of general skills not found in a specific career-oriented program. Students who study the liberal arts and sciences at Reed acquire an array of experiences, perspectives, and interpretive methods with which they can tackle any problem, within any career." The college provides connections to many internship, summer study, and undergraduate research opportunities that facilitate students' career search and development.

RICE UNIVERSITY

P.O. Box 1892
Houston, TX 77252-1892

(800) 527-OWLS/(713) 527-4036

www.rice.edu

Number of undergraduates: 2,700
Total number of students: 4,285
Ratio of male to female: 52/48
Tuition and fees for 1999–2000: $15,796
% of students receiving financial aid: 85
% of students graduating within five years: 86
Endowment: $2,790,627,000
Endowment per student: $651,255

Overall Features

Rice is one of the newer members of the group of America's elite universities. William Rice, a transplanted New Englander who made his fortune in Houston, donated the funds in 1891 to endow a college modeled after a combination of Princeton and Oxford with an emphasis on the liberal arts curriculum in a residential living environment. The first class entered in 1912. Thanks to William Rice's generosity the university created a huge endowment over time, which explains its relatively small student population for such a top-flight teaching and research university. It ranks second only to Princeton in its endowment resources for each enrolled student. The tuition is the lowest in the nation of all the top-rated independent universities. Rice has deliberately maintained a two-to-one ratio of undergraduates to graduate students to keep the teaching focus on the former. The small overall size and low student-to-faculty ratio have created an exceptional institution, one that can retain a world-class research faculty that is also dedicated to teaching and mentoring undergraduates through small classes and research and internship opportunities. With a large faculty of 460 full-time teachers, Rice is able to structure its curriculum into six distinct undergraduate colleges: three divisions in the liberal arts (humanities, social sciences, arts) and three professional programs (art and architecture, music, and natural sciences/engineering). The professional schools have long enjoyed a top reputation because of the small size of the programs and the excellent faculty and facilities. The humanities program is expanding in strength and size to match the opportunities of the other programs.

The student body is one of the most selective and talented cohort of men and women to be found in any of the major universities. Of the first-year class, 26 percent are National Merit Scholars, 88 percent graduated in the top 5 percent of their high school classes, and their average admission test scores placed them in the top 10 percent of national test takers. Thanks to its endowed funds, Rice can offer admission to any student who qualifies irrespective of need. Similar to several of the Ivy universities and Stanford, Rice has reconfigured its aid package to lower the portion of educational loans students must assume. It may have gone one step farther by guaranteeing scholarship recipients that their loans will not exceed $2,450 each year of study. The generous aid program combines with the low tuition to make Rice one of the most attractive institutions for outstanding students of all backgrounds and means. Over one-third of the student body is of color, and all fifty states and seventy foreign countries are represented. Over one-half of the students are still from Texas, however. The admissions committee is working to widen the geographic representation to bring still further diversity to the campus.

The atmosphere is a serious one. Students enter Rice to avail themselves of the outstanding programs of study and renowned faculty, 98 percent of whom hold a Ph.D. in their respective disciplines. A majority of undergraduates major in engineering, natural sciences, and the humanities, in that order. The music and architecture schools purposely remain very small and highly selective. An honor code guides the academic life of the school. Students take exams free of proctors and are expected to present only their own work. Violators must come before a student judicial board. The community treats the honor system with great respect, according to students and administrators.

The residential system is unique among American universities, with the exception of Harvard and Yale, since it was modeled on Oxford University. The university believes this is the heart of the Rice experience for its undergraduates. There are eight residential colleges or clusters, each of which houses 225 men and women. Each cluster has its own dining halls and social commons room. A faculty member or master is in residence and helps set the tone and activities of the cluster, together with the students who govern most of the life of their college. The particular college a student experiences will play a major role in defining his or her growth and development socially and intellectually. Each cluster supports intramural athletics and social events. Students are assigned to their cluster for all four years as a means of forming strong ties with their peers. The only problem with this living plan is the fact that at present only 65 percent of the students can be given housing. There is serious discussion of building a ninth house to guarantee all students the opportunity to experience this special living arrangement.

Rice participates in intercollegiate athletics at a highly competitive level in Division I-A. Many students engage in intramural sports or in the large number of clubs and community service organizations that are supported by the university. There are no fraternities or sororities on campus because of the house system. Rice is a handsome campus four miles from the center of Houston, which is now the fourth-largest city in the country and growing at an accelerating rate. This is fueled by the technology, aerospace, and business industries located in the area. Rice's faculty and student body benefit from the opportunities for internships, research, and funding these groups provide for career and academic projects.

What the College Stands For

Rice's educational leaders believe they have not wavered from the original vision and mission to combine the finest quality teaching of highly motivated and intellectually curious young men and women with significant areas of research, all to the ultimate purpose of serving the public good. The concept that scholars who are engaged in research and experimentation in their discipline are the most exciting teachers drives the size, curriculum, and modes of learning at Rice. Bringing the methodology of scholarship and discovery into the undergraduate classroom will enhance the skills and interests of their students. The public will then be served by the new knowledge that will result from the facilities and support with which the university provides its faculty, and the education students receive will ready them for positions of leadership and innovation in both the public and private sectors. To emphasize the point, two of the faculty in chemistry who have been honored with Nobel prizes have taught first-year chemistry. Their work was recognized for its contribution to society. How can a budding scientist not be enthralled to interact with teachers of such prominence? As the university states, "In a good university, the traditional functions of teaching, research, and public service can scarcely be separated—effective teaching demands continual research, and the development of the intellectual life of young people is a major public service." The guiding spirit is the value of arming students with a combination of a broadly based education with advanced study in a chosen field of specialization. Engineers and scientists should become versed in the great literature and thought of Western civilization, which will guide them as technical leaders in society. Those who will engage in commerce or law or education should be familiar with the fundamental principles of science and mathematics so that they can engage in intelligent dialogue with the scientific community. Certainly Rice's generous financial aid program ensures that any student with a serious commitment to learning will have the opportunity to receive this first-level training. The university sees itself as an agile community of teachers, researchers, and learners that can adapt to new frontiers of knowledge because of its financial and human resources, its small size, and its commitment to serve the nation.

Curriculum, Academic Life, and Unique Programs of Study

We call attention again to the particular strengths of the academic program: six distinct colleges of study within a small undergraduate college and a smaller graduate school. Historically, engineering and the natural sciences have enrolled the largest number of students. The colleges of the social sciences and humanities are growing in enrollment, while those of architecture and music are quite limited in numbers. Whatever field of concentration a student opts for, he or she will be required to take foundation courses in the broad spectrum of arts and sciences before moving on to the field of concentration. As the school notes, "The academic philosophy at Rice is to offer students beginning their college studies both a grounding in the broad fields of general knowledge and the chance to concentrate on very specific academic and research interests. By completing the required distribution courses, all students gain an understanding of the literature, arts, and philosophy essential to any civilization; a broad historical introduction to thought about human society; and a basic familiarity with the scientific principles underlying physics, chemistry, and mathematics. Building on this firm foundation, students then concentrate on studies in their major areas of interest."

Rice prides itself on being on the cutting edge of research and teaching in chemistry, physics, and engineering. Some of the specialized fields of study offered are biochemistry/cell biology, bioengineering, neuroscience, and space physics. Opportunities abound to develop combined majors of one's choosing; it is not unusual to major in a science and minor in a foreign language or religion and philosophy, for instance. The faculty encourages students to engage with them in independent learning projects or scientific research. The joining of intelligent and goal-oriented students with a world-class faculty teaching small classes creates a serious and demanding learning environment at Rice. Most teachers grade on a curve system because of the talent pool in their classes. This has the unpleasant effect of setting up a competitive edge. Also unique for a small undergraduate institution is the presence of several centers for specialized studies, such as the Institute for Culture and Languages, an interdisciplinary program designed to create global awareness and broaden cultural studies; the Baker Institute for Public Policy; and the partnership program with Bremen International University, the first independent university in Germany, which will admit its first class of students from all corners of the world in 2001. Rice is one of the leading supporters of Bremen and will encourage its students to study in Bremen's school of engineering, sciences, humanities, or social sciences with the international perspective that is emphasized. Rice is also expanding its own international studies programs for its undergraduates in France and other countries following its vision of preparing leaders in all fields for the global economy and intercultural world in which they will work.

Major Admissions Criteria

More than 6,000 high school seniors apply each year for a first-year class of only 650 students. The dean of admissions states that two sets of questions are asked that will determine which candidates will be selected from such a large and qualified pool:

- When you are exposed to intellectual opportunities, how do you respond? Have you taken the most challenging courses offered in your school or reached out to enhance your academic interests in other ways?

- How have you performed in a demanding learning environment? Do your grade performance and teacher comments indicate a true desire to learn, to make demands on your intellectual abilities and extend your knowledge to the maximum?

The committee must then choose those students who are most likely to succeed in the particular college for which they have applied. Musicians, artists, and architects must demonstrate through portfolios and performance certain talents and strengths, in addition to intellectual ability, while future engineers need to show mathematical and science aptitude and course foundation of a high caliber. Standardized test scores are not the primary factor in the decision process, but they are considered as a partial reflection of academic preparation and ability. In part they are a necessary screening device. The ultimate question asked of each applicant is how he or she has acted on his or her interests and abilities—this will determine who will best use the many advantages Rice provides. Because of the university's commitment to educate young men and women of many disadvantaged backgrounds, the admissions committee gives weight to applicants of color and ethnicity. However, racial quotas and special preference cannot be used in Texas to admit students, as a result of a recent Federal District Court decision. The committee tries to ascertain the backgrounds of its candidates through the important personal statements required on the application and through personal recommendations from school and outside supporters. Rice is determined to keep its doors open to deserving students who can take advantage of its learning resources.

The Ideal Student

Do not consider Rice if you are not serious about studying and learning. These activities come first for all students on campus. Social life is squeezed into a demanding and busy daily life. Due to the university's outstanding science and engineering departments and the professional goals of a majority of undergraduates, the overriding tone of the community is conservative to moderate, one that does not take much time to consider the politically correct issues and activist behavior that define many of the other top colleges and universities. It helps greatly for students to be focused on a field of interest, to have a desire to work hard, and to enjoy closely interactive teaching. The sciences and engineering programs are the dominant ones on campus, while a small cluster of students will be highly focused on their professional studies in art and architecture or music. There is not a great deal of casual course selection or a laid-back attitude toward studies. Extracurricular activities are plentiful and can be fitted into one's schedule comfortably. "Traditional" is the operative word to describe the attitudes and lifestyle of the majority of Rice students. There is a Southwestern flavor to the campus because of the location and the dominant percentage of Texans in the student body. Jeans and boots will be more acceptable than Birkenstocks in this setting.

Student Perspectives on Their Experience

Students choose Rice for its academic programs, faculty, college system, and size. "The quality of the architecture school was my primary factor," says a sophomore, "however, the social scene created by the college system tipped the scales in favor of Rice over other schools." A senior lists "no Greek system, the college system, and academic reputation" as her primary reasons for selecting Rice. Rice is praised for its provision of individual attention to students and its focus on undergraduates. "The college system creates a grouping of relatively small, tight-knit communities that provide many friendships and possibilities for involvement but also foster a sense of belonging and pride."

Those who do best at Rice are "the people who see that Rice offers more than academics and enrich themselves in the many diverse opportunities for involvement available," according to a sophomore. Student success is assisted by "the support of peers, the reasonable guidance of professors, and the ability to remain calm and rational when faced with deadlines and stress." Another student identifies "peer help and individual initiative" as important.

Diversity at Rice is promoted by student organizations, recruitment and retention of a talented and diverse student body, and the residential college system. "Many different clubs exist which act to strengthen bonds within certain groups," says one student, "as well as sponsor activities to celebrate aspects of the group with the entire university." The university benefits from "the communities created by the college system and the opportunities, relationships, and unity therein."

What Happens After College

The university boasts that over 70 percent of its graduating seniors in any given year are admitted to their top choice of graduate school. This is entirely believable given the selectivity of the students in the first place and the rigorous training they will receive over their four years of study. The largest number of graduates continue their studies in engineering, followed by the natural sciences, business administration, and architecture. Rice graduates are awarded a large number of National Science Foundation Fellowships in science and engineering. Premedical concentrators also have great success in admittance to medical schools. Similar to the students in the majority of other Hidden Ivies, Rice students succeed because of their close relationships with their teachers and the chance to do research projects or independent work under their aegis. The bottom line is that if one does well in one's undergraduate studies, the opportunity for an advanced degree at a top graduate school will follow naturally. The university is also aware that many graduates want to work upon graduating. A comprehensive career advising and placement office is in place; there is a proactive approach to encouraging major companies to recruit on campus or over the Internet.

SMITH COLLEGE

7 College Lane
Northampton, MA 01063

(413) 585-2500

www.smith.edu

> Number of undergraduates: 2,500
> Total number of students: 2,900
> Ratio of male to female: 100% female
> Tuition and fees for 1999–2000: $22,722
> % of students receiving financial aid: 58
> % of students graduating within five years: 81
> Endowment: $885,000,000
> Endowment per student: $305,172

Overall Features

Smith takes great pride in its long and distinguished history as the largest of the women's colleges in America. Sophia Smith founded the college in Northampton, Massachusetts, in 1875 to provide intelligent, motivated young women with a liberal arts education the equal of that to be found at neighboring Amherst, Williams, Dartmouth, and Yale. The college has maintained its high academic standards and expectations for students but with greater flexibility for choosing a field of study, undertaking research projects and internships, and experiencing academic and social life on a number of other select campuses or abroad. Like its peer institutions of note, Smith has allowed far greater social freedom as well. Students live in one of thirty-five residential houses of a wide variety of styles and sizes. They accommodate anywhere from 10 to 100 women who are responsible for governing themselves and planning activities. All four classes, from freshman to senior, are mixed in the majority of the houses to provide more interaction among the students.

Like many of the other top liberal arts colleges, Smith governs its academic behavior by an honor code that appears to influence the students beyond the classroom. Smith women are self-governing and serious about learning and creating their own lifestyles. They pick up early in their college careers the Smith ethos of becoming independent thinkers in and out of the classroom as a preparation for a life of successful careerism and community service. The college leaders celebrate actively those alumnae who have achieved success in the professions and business world and the arts. Smith women are liberal and outspoken in their views, hardworking, and assertive in accomplishing the goals they set for themselves. Women's issues are an active subject of dialogue on campus and in the classroom. Courses on feminist thinking, women's literature, and history are offered, and numerous speakers are invited to campus to discuss these topics. A great many Smith students are very talented in or committed to the arts; many are competitive athletes. The faculty is excellent, with 96 percent holding an earned doctorate in their discipline. All faculty members are expected to teach and to keep up on research in their field, and the college notes with pride that the faculty won six research fellowships in 1999 from the National Endowment for the Humanities, the most of any college. Smith can afford a large faculty, some 265 full-time professors who teach in thirty-seven academic departments. The faculty-to-student ratio is ten to one, with many classes averaging five students. Students have 1,000 courses to choose from in fifty different areas of study.

The Smith campus of 125 acres is very attractive, and its facilities for academic programs, athletics, and the arts exceptional. The heart of the campus is the library, which contains 1.2 million items, making it the largest undergraduate liberal arts library in the country. The performing arts center is huge and technically very modern, allowing for every form of visual and performing arts study and practice. The science center is also up-to-date and addresses the needs of a strong science department and the many women who concentrate their studies in the sciences. There is even a separate science library of distinction. An art museum and collection of some note is used by humanities and arts majors particularly. The college sponsors fourteen varsity sports for intercollegiate competition, including a major crew program—Smith has its own lake and boathouse on campus. There are indoor and outdoor riding rings for equestrian competition and a large indoor athletic complex. Smith is determined to be in the forefront of learning technology by offering a large number of computer facilities, training, and access for all students on a fully wired campus.

The president of Smith, Ruth Simmons, is an African American who wants to bring more multicultural diversity to this formerly all-white, upper-middle-class community. She is the third woman president in a row for Smith, which has chosen female presidents since 1975. To attract young women of all socioeconomic and racial backgrounds, the college provides financial aid for 58 percent of those who demonstrate need. The minority population presently is 20 percent of the community. The average grant is very generous and well above average for the majority of independent liberal arts colleges. How does a college of only 2,500 students make such physical, financial, academic, and human resources available? The answer is endowment. Smith's invested assets stand at $885 million, which puts it in the rarefied company of the richest coeducational colleges in the nation. The traditional notion that women will not be generous givers to their alma mater is simply not true. This past year Smith garnered $53 million in gifts from alumnae, friends of the college, and parents.

What the College Stands For

Smith seeks to educate its students in the foundations of the liberal arts and to prepare them for active careers that challenge the boundaries in "traditionally male" fields. As one administrator puts it, the educational goals for Smith students are "to develop their leadership and other intellectual and civic virtues; to produce highly able and articulate graduates who are equipped in the best fashion of the liberal arts to pursue their vocations and other life goals; and to prepare women for career and other objectives, regardless of whether these have traditionally been regarded as 'male.'" Another senior official notes that "new programs we have instituted or are instituting, specifically those that emphasize writing, speaking, and quantitative skill building across the curriculum, will build on our traditional strengths by ensuring that all Smith graduates write cogently and persuasively; speak publicly with confidence and coherence; and answer complex questions through the careful analysis of quantitative data. Because of our new universal internship program, which is designed to bridge theoretical knowledge to the practical application of problem-solving in a variety of work environments, Smith students will graduate with considerable work experience and the demonstrated ability to interact effectively in a professional environment, the personal resourcefulness to provide solutions to problems in organizations, and the ability to connect theory and practice. Finally, Smith has long had a commitment to undergraduate research, and with the recent founding of the Kahn Liberal Arts Institute, which will provide student fellows with the opportunity to conduct multidisciplinary research on a series of fascinating questions, Smith graduates will have considerable experience in using the tools of scholarship to answer vexing questions, with the additional benefit of having analyzed problems from the multiple vantage points that interdisciplinary research allows."

Smith sees student success as being helped by the college's ethos of educating young women to achieve their best in any field of endeavor, the community environment of the college, the faculty, and the school's strong alumnae network. "The very culture of Smith breeds success," says an administrator. "The stories of alumnae who have done extraordinary things are told and retold, and those alumnae act as mentors to current students (the 'old girl' network rivals anything the 'old boys' can come up with at other institutions). Smith celebrates academic success (in comparison to other places where athletic and social attributes often receive top billing). The housing system, small and more home-like than traditional dorms, provides a support system for students to excel. Smith has an extraordinary teaching faculty, who know students individually and are consistently demanding. And there are class deans and residential life staff who keep individuals from falling through the cracks."

As is obvious from these statements, Smith, like its single-sex peers, sees in its all-female environment a very particular mission toward educating and instilling a high degree of self-confidence in young women. Students are helped, says one official, by "the belief from the first day of the first year that anything is possible. That is inculcated from the beginning."

Curriculum, Academic Life, and Unique Programs of Study

With its broad, sweeping curricular offerings, Smith gives women considerable freedom to choose what they wish to study. There are few distribution requirements and a multitude of courses to fulfill them. The only universal requirement is an intensive writing course. All other requirements are determined by the departments for major and minor concentrators. Students bear the major onus for deciding which of the fifty subject areas they will commit to. The faculty encourages building a major and minor curriculum or selecting one of the many interdepartmental majors. All students are encouraged to engage in research or independent study projects. Smith is probably the first college of note to have created the concept of junior year abroad study for its students. Today a majority of women will take advantage of the college's own programs in Florence, Geneva, Hamburg, and Paris or those sponsored by other major colleges.

Concentrators in all fields of study are encouraged to broaden their intellectual and social experience either by studying abroad or by participating in an exchange year on the Twelve College Program, the Pomona College Exchange, or the Historically Black Colleges Program. One-half of the student body will gain new experiences and learning through the elaborate internship opportunities Smith sponsors directly, and all students are guaranteed an internship if they so desire. These include the Semester-in-Washington Program and the Smithsonian Institute Internship. A newly created operation run by the Career Opportunities Office called PRAXIS provides opportunities and funding for a wide variety of summer internships for the purpose of combining theory and classroom learning with actual exposure in the field. Other interesting intellectual pursuits are available throughout the college, including a new scholarly journal founded at Smith in conjunction with Wesleyan University, *Meridians: Feminism, Race, Transnationalism*, focusing on women of color throughout the world.

The academic schedule is divided into two semesters of thirteen weeks each and an interterm of three weeks. Students can immerse themselves in an independent project or internship that usually relates to their academic concentration. Still another exceptional opportunity for Smith women is membership in the Five College Group centered in the Amherst area. The five colleges sponsor student and faculty exchanges, joint faculty appointments, and courses in order to enlarge the course options for students. In a typical year 800 Smith women enroll in courses at the other colleges. Not to be taken lightly, social life is also enhanced by the many exchange programs Smith sponsors. The newest academic addition to the curriculum is the first engineering major in a women's liberal arts college. Majors in electrical, computer, and environmental engineering are now available to Smith undergraduates. The college is well on its way to securing the $250 million necessary to fund this new department of the college. Engineering majors can elect to spend their junior year at the Thayer School of Engineering at Dartmouth. A combined 3/2 Bachelor of Science in Engineering degree is also possible in conjunction with Dartmouth and the University of Massachusetts.

Major Admissions Criteria

The admissions office says, "We are looking for students who have challenged themselves by taking a broad, rigorous college prep program in high school, students who are bright, motivated, and interested in actively participating in their educational experience." Top admissions criteria are identified as:

1. High school performance for four years—level of rigor, courses selected and performance.

2. Standardized testing.

3. Letters of reference.

Another official lists "academic merit/ achievement, community enrichment, and leadership potential." And another, "aptitude, leadership, and achievement in one or more areas that sets students apart."

The Ideal Student

Choosing Smith College should not be based primarily on the fact that it is a women's college. This is secondary to the opportunity Smith provides to experience one of the great educations available in a residential liberal arts college. The faculty, the curriculum, the resources, and the many diverse opportunities for intellectual and personal growth should appeal to a self-motivated, independent young woman who enjoys the company of other intelligent, curious, ambitious, and independent thinking women. The Smith student of today is not a traditional white female with good brains and good manners who is dependent upon the elite men's colleges of the past for her social life. Today's Smithie is liberal-minded, assertive in espousing her point of view, likely to be on scholarship or of a nontraditional background, and welcoming of her peers, who may have a different social or cultural orientation from her own. Most of all, she is dedicated to acquiring strong intellectual training and preparedness for the competitive world after college. She is also likely to engage in contemporary issues of concern such as social injustice, women's rights and privileges, and equal treatment whatever one's sexual habits and persuasion. As one administrator puts it, Smith looks for "very bright students who want to develop their strengths as women; activist types; and science fans." Academic conservatives or social wallflowers who do not enjoy an intellectual discussion or heated debate of sensitive social issues will not feel at home on the Smith campus. Students who do best at Smith are those who "will feel comfortable in a community of individuals; students who will speak their mind but listen with care and respect to opposite views and beliefs; and students who are compassionate, studious, positive." Another administrator says, "Strong-minded, competitive, courageous women do best here. We want bright, fully dimensional students who want to put their passions to work."

Student Perspectives on Their Experience

The college administration recently surveyed the student body regarding its satisfaction with life at Smith, and nine out of ten seniors expressed satisfaction with their overall experience. Eighty-five percent affirmed that the academic training had increased their ability to write effectively, think analytically and logically, understand themselves, synthesize ideas, and gain knowledge of their particular academic field of study. These are exactly the skills that liberal arts colleges set as their goal. The retention rate for undergraduates at Smith is well above the national average, but not as impressive as a number of the other select colleges. The cause of this is not entirely clear, but each year there are a number of women who transfer to larger, coed, urban universities. Several transfer students with whom we talked expressed great respect for the academic programs and the faculty they experienced for one or two years at Smith, but they felt they were ready for a larger, diverse, coed community in a setting with more action than Northampton could provide.

Smith has changed, however, as one graduate and current administrator notes: "Smith seems to be a more diverse and vibrant place than when I was here in the eighties. There is a sense of pride more loudly proclaimed on campus. More risks are being taken and with great effect—no stasis here." Current students choose Smith because of its academic programs, reputation, five-college consortium, alumnae network, personality, and financial resources. As one student notes, "I applied to Truman State in Missouri and to Smith College. Smith was my 'dream' school. A dream because of its reputation and cost. I was accepted to both and because of Smith's need-based financial aid, both colleges were almost the exact same in cost."

And while many students did not initially think they would attend a women's college, visits to Smith and discussions with Smith representatives and alumnae showed them the reality of life on campus today. "I heard about Smith at a college fair. The representative had a good handshake, she looked me in the eye and gave me a lot of impressive information in a short amount of time. I was very impressed with Smith's mailings, which were frequent and informative. I also liked how the students that were pictured were not all skinny, blond, and white. It appeared to be an open environment, where all walks of life were accepted . . . I also fell in love with the campus. I am a big one for wide-open spaces, and Paradise Pond and the Athletic Fields are beautiful! The softball field is the best I have ever played on—the grounds crew is amazing. The housing system also caught my eye. Each house, for the most part, has its own kitchen. This means you can roll down to breakfast in your pj's and nobody cares . . . The networking that Smith was known for also influenced my decision. I don't know what I want to do still, but I do know that somewhere in this world there is a Smithie who can help me get my foot in the door when I do figure it out."

A junior says: "I chose Smith because I was looking for a women's college, and early in the college search process Smith emerged as one of the leaders. At the time, I planned to be a women's studies major, and the obvious place to do that was one where women were put first and foremost. Since then, I have changed my academic course and I've been glad that I ended up at Smith, for very different reasons than I originally chose. I first toured Smith during a summer and though I was impressed, I wasn't totally convinced that it was the right place for me. That became quite clear though when I visited as a prospective student, just days before I had to choose a college. There was such energy there among the students on that second visit. Ironically, the thing that really sold me on Smith was the swim team. I thought of myself as a student first and an athlete second, but I knew that all the schools I was considering were academically strong. Smith distinguished itself as the socially superior school as well. As I attended classes and sat in the dining room and wandered around campus, I met people that I knew would be my friends if I gave them the opportunity. It was then that I realized Smith was perfect for me. Since then, I have sometimes wondered what life would have been like had I chosen another women's college. Though it is conceivable I could be as happy somewhere else, it's simply not possible that I could be one bit happier. I love Smith and have never regretted my decision."

Students speak very positively about Smith's professors, classes, and their peers, and about the significant amount of work they must do. However, they feel welcome, challenged, and supported on campus. "I think Smith does an excellent job of teaching us how to communicate. It's great to have an idea or opinion, but if you can't express it verbally or through written words then it is wasted. Smith uses the presentation of papers to increase public speaking skills, and most professors demand a lot out of any paper. And man do we write a lot. I think you could get just as good an education at most colleges, if you applied yourself. However, at Smith you have to apply yourself even if you are wicked smart. Class participation is a must to receive

an A in most courses because class size is so small. Smith also teaches us how to think and to continue learning throughout our lives."

Another student notes, "Smith professors and staff provide a sturdy and extensive support network for students. They are watchful of more than just our academic success; they are concerned also with our emotional, social, and spiritual well-being. Without being intrusive, they nonetheless are involved in our lives, by our decision. Few schools can boast such close relationships and important bonds between the students and the faculty and staff." The five-college program works for students, even though it is not completely convenient to travel the fifteen to twenty minutes from one campus to the next. Says one student, "If Smith were not part of the Five Colleges, it would be a very different place. In my opinion, one of Smith's greatest strengths is its ability to fully exploit all the resources offered by the other area schools without becoming dependent upon them. I thought Five-College cooperation wasn't really as important as the admissions office made it out to be. Now, I am so happy that I was wrong. Smith is an incredible school independently, but the combination of UMass Amherst, Amherst College, Hampshire College, Mount Holyoke College, and Smith is ideal. You can have the best of all worlds . . . namely, you can have all Smith offers and still have all the benefits of the other schools without making any sacrifices."

Students who do best are "those who get involved, meet people, develop relationships with professors and administrators, apply themselves academically . . . those who are able to maintain a balance between interests, duties, wants, and playtime. Balance is key, but talking with your professors also helps. If they see you taking an interest, things just seem to run smoother. The Jacobson Center is available for peer editing and tutoring. There is a campus chapel for the spiritual side, and nature surrounds Smith. I think that if you look hard enough in yourself and decide what is truly important to you, you will find others with the same interests and form friendships and support groups for sharing ideas, plans, hopes and fears. But you do have to find a release somehow. You can't just study all the time." And from a junior, "The students who do best at Smith are those who put themselves to work seriously at getting the most from academic opportunities without sacrificing the social component of being a college student. At Smith, with some of the best professors in the country and facilities to match, it's all too easy to become concerned only with book learning. But Smith has a lot more to offer than that . . . namely, some of the most fascinating young women in the country."

What Happens After College

Confirmation of the ambitious nature of Smith women and the outstanding education they receive is the track record of acceptance to top graduate schools and the large number of highly competitive fellowships they win. Over the past decade these include twenty Fulbright, thirteen National Science Foundation, four Truman, two Mellon, and one Rhodes Scholarship, plus a host of other prestigious awards for study and travel at home and abroad. Typically two-thirds of a graduating class go directly to work in a variety of professions, and another 25 percent enter graduate school. Smith estimates that over time 90 percent of alumnae will pursue advanced studies. Smith has cherished and nourished its old-girl network that assists present and recent graduates to obtain jobs and internships in their chosen profession. Many Smith alums have told us of their positive experience in this regard. As one administrator puts it, Smith graduates are "prepared to shift with nimble grace to meet all challenges."

SWARTHMORE COLLEGE

500 College Avenue
Swarthmore, PA 19081-1397

(610) 328-8300

www.swarthmore.edu

Number of undergraduates: 1,370
Total number of students: 1,370
Ratio of male to female: 47/53
Tuition and fees for 1999–2000: $24,190
% of students receiving financial aid: 48
% of students graduating within five years: 91
Endowment: $834,000,000
Endowment per student: $608,759

Overall Features

Swarthmore, one of the group of liberal arts colleges founded by the Society of Friends, commonly referred to as the Quakers, is distinguished both for its academic programs and quality of student body and for the spiritual and ethical values that define the community. Thanks to its reputation for academic excellence, the quality of its faculty and their commitment to teaching, and its enviable financial strength, Swarthmore has become one of the most selective institutions in the nation. Its standards for admission are equal to those of the Ivy colleges by every measure. Only 20 percent of candidates are offered admission each year. The size of the endowment and the ratio of financial resources for educating each student is a story in itself. Swarthmore can afford to attract and retain academic stars in their individual disciplines

who combine research and writing with teaching one of the brightest and most intellectually committed group of students to be found on any campus. The student-to-faculty ratio is eight to one, which allows for genuine personal interaction and intellectual exchange. The college is also able to meet the financial need of every student who qualifies for admission. Here is one of the handful of colleges that can still maintain a need-blind admission policy. Eugene Lang, a graduate of the college who used his earned wealth to create the I Have A Dream Foundation, recently donated $30 million for scholarships to ensure the continued presence of disadvantaged students on campus. Another famous former scholarship student, James Michener, also left millions to the college, part of which was designated to fund financial aid. This is how a very small college can attract and enroll a diverse and talented entering class each year.

Students come from all fifty states and thirty-nine foreign countries, with the Middle Atlantic states having the largest representation. At present almost one-half of the student body is of color and one-half of all students receive financial support from the college.

Swarthmore students are renowned for their academic work ethic and genuine interest in intellectual pursuits. As with the other outstanding strictly undergraduate colleges in this study, all of the campus resources are directed toward the undergraduates, and the faculty teach a full course load and are available and helpful to students. At the same time, they place great demands upon their students, as their goal is to teach them to think for themselves in a critical, analytical way, in part by assigning a heavy reading load and requiring independent study and research assignments. Classes are small and depend on student participation at all times. Swarthmore is justly proud of its magnificent campus, which is a 330-acre, nationally designated arboretum. Philadelphia is a short thirty-minute train ride away from campus for a touch of reality when desired. Swarthmore's historical affiliations with its neighboring colleges, Haverford and Bryn Mawr, add to the intellectual and social opportunities for undergraduates. In a supposedly non-sports environment, Swarthmore fields twenty-two varsity athletic teams for men and women in Division III competition. There are numerous intramural activities for the remaining students. It is safe to say that the typical Swarthmorian puts academics well above athletics in importance but does enjoy the chance to get away from his or her studies through sports and the multitude of organizations available on campus.

What the College Stands For

Although Swarthmore considers itself a secular institution today, its mission statement clearly reveals its commitment to Quaker ideals: "We seek to illuminate the lives of our students with the spiritual principles of the Society of Friends." It adheres to the Quaker ideals of "hard work, simple living, and generous giving; personal integrity, social justice, and the peaceful settlement of disputes." Simplicity in all aspects of life is another significant tenet encouraged among students. This set of ethical values permeates the campus and the classroom. Students certainly adhere both to the work ethic and respect for one another's opinions and beliefs. A key element of the Quaker heritage is a commitment to community service, which all students are encouraged to participate in through the Swarthmore Foundation, which sponsors students in a wide range of service activities. There is a definite liberal attitude toward social and political topics but less of the aggressive activism and effort to effect conformity of thinking that prevails at other well-known liberal-thinking institutions.

What the college sees as important in itself encompasses "very committed professors who are as devoted to teaching as to scholarship. Swarthmore has one of the lowest student-faculty ratios in the nation (if not the lowest). A huge endowment that lets low-income students concentrate on their studies and not their part-time jobs . . . A campus 'culture' in which academic competition—believe it or not—barely exists. Grades are never posted. The emphasis is on one's personal best." As another administrator puts it, "A key objective here is to teach analytic ability—the ability to see beyond simplistic, black-and-white arguments—and to impart what our president, Alfred H. Bloom, calls ethical intelligence . . . The small class sizes and close contact with professors are essential. Also, our sole focus on the undergraduate ensures research and other opportunities that would otherwise be the province of grad students. The small size fosters an intimate community that enriches the college experience for our students."

Curriculum, Academic Life, and Unique Programs of Study

In keeping with its mission of exposing its students to a broad range of history, science, human behavior, creative expression, and languages, Swarthmore requires a number of courses across these disciplines. However, there is no core curriculum and the choice of subjects to meet the distribution requirement is extensive. Swarthmore's curriculum and academic opportunities have broadened over time to allow students many options on and off campus. All first-year courses are graded on a pass/fail basis to alleviate the academic pressure and potential for competition within the class. Quite unique for so small a liberal arts college is the presence of the school of engineering, available to those who plan scientific training and a career. There are a number of cooperative programs for advanced study with leading university graduate engineering schools. Serious students are encouraged to combine fields of study, assist faculty in research projects, study off-campus in meaningful programs, and take advantage of the academic resources of Bryn Mawr, Haverford, and the University of Pennsylvania.

Programs in interdisciplinary fields cover a diverse range of topics, from environmental to Francophone or German to peace and conflict to public policy or women's studies. For the highly motivated, the college has created an external examination program that lets students participate in eight honors seminars in their junior and senior years. They are examined at the end of their special studies by a board of external examiners from other colleges and universities. An unusual policy that helps to explain the accomplishments of the faculty is the four-year cycle of sabbaticals in order to carry on teachers' research, travel, and writing. An administrator says, "What strikes me here is the integrity of the academics. Swarthmore has created a community where learning is very highly valued. I'm continually impressed by the students. Unlike those at some schools, Swarthmore students are not obsessed with GPAs and high-paying careers. Nor is the experience here about high-profile sports or partying. To me, it's college in its purest form. They're here for an intense intellectual experience."

Major Admissions Criteria

What does selectivity translate into, in terms of the credentials for admission to Swarthmore? Only 36 percent of high school valedictorians and salutatorians, 28 percent of applicants who ranked in the top 2 percent, and 21 percent in the top decile who applied for the class of 2003 were offered admission. The average SAT I scores were well into the 1,400s combined. The college takes seriously its policy of enrolling serious learners who demonstrate intellectual talent and the commitment and initiative to take advantage of the college's resources. Its need-blind policy is demonstrated by the awarding of financial aid for some candidates who need a full subsidy, to the tune of $32,000, in order to enroll.

The Ideal Student

Swarthmore students put learning and intellectual exploration above all other priorities. They are attracted to this small, intense college for the quality of its faculty, its resources for learning, and probably most of all, the peer group they will interact with for four years. The ideals of respect for others and a determination to take advantage of their education lead high-performing high school students to choose Swarthmore over the comparable leading colleges they would qualify for. The non-academic features of the bigger, more traditional colleges do not appeal to them as much as do the premier teaching and learning that distinguish Swarthmore. So a potential student has to be clear about his or her own priorities and how hard he or she is willing to work to have a successful experience. Students are expected to be independent, to be self-starters open to new and differing ideas. Passion for an academic subject or the arts or athletics or helping others is a trademark of most of the students.

Student Perspectives on Their Experience

Location, academics, and the personality, philosophy, and feel of the college attract prospective students to Swarthmore. Says a junior, "I chose Swarthmore mainly because the college simply felt right when I visited it as a prospective student. The combination of strong academics and social life fits me, and the location is great. The campus is beautiful, being on the Scott Arboretum and with the Crum Woods adjacent to the campus (a great place to escape to, to go running in, and be 'off-campus' while really not going far on a busy day). The suburban setting is also nice, because students are not limited to the college campus as a safe haven. The SEPTA station (local commuter rail) is literally on the edge of campus, there is a mall within walking distance, and the town of Swarthmore—'the Ville'—is all around. The main quality that first interested me in Swarthmore was its reputation for a strong liberal arts education at a small college in a suburban setting. Visiting the campus was what set it apart from the other similar schools." A recent graduate describes the differences she liked about Swarthmore when she visited campus. "First, I was struck by the college's incredible beauty; second, and much more important, I noticed the obvious intellectual atmosphere, different from that of other schools because it appeared to be one in which students did not take themselves too seriously. They seemed excited and genuinely interested in learning for its own sake. They wanted things to be done in their community or the world, not for personal rewards or recognition. The students felt their efforts were worthwhile and their goals should be attained whether they or someone else was responsible for them. It appeared to be a very sincere, grounded, yet exciting atmosphere (and since attending, I have not been disappointed)."

According to students, what Swarthmore

does well is teach, encourage, challenge, and support. "Swarthmore is excellent in inculcating academic rigor and curiosity in its students. It therefore graduates very intellectual and maybe rather idealistic students," says a senior. The workload is rigorous, and students will spend a lot of time on their own and with others pursuing their intellectual interests. "Students who do best here must be prepared to study *all* the time. Grade inflation does not exist and so to get an A one has to spend about ninety percent of the semester in the library. That is the only way to stay on top of the curve. Therefore students tend to be very academic and have little time to do co-op programs to gain industrial experience alongside their coursework during the semester. Students succeed here because of the elaborate support systems in place to aid in academic advising, tutoring, mentoring, etc. Students feel welcome to go to any source of support when they need it so it is very easy to find help here."

A junior says, "Swarthmore has a strong sense of community. Perhaps that comes from being a small school, but it is something that I really like. There is an incredible amount of personal attention given to each student, and it isn't uncommon for a professor to become a friend as well as an educator. Students are often on a first-name basis with professors, and learning becomes a mutual experience. Students at Swarthmore have a big input into the learning that they receive—be it with classroom participation, research alongside a professor, or in conjunction with the administration. Swarthmore is a great place to get a solid broad-based liberal arts education—not just superficial in a lot of areas but an under-

standing of each field. Students have widespread talents and a lot of the time you will find people with very diverse interests or with interest combinations that are unexpected. I chose between French and chemistry for my major—and although I declared French, I still study chemistry. I am working in a biochemical engineering and testing lab this summer—following a semester abroad in Grenoble, France, on a Swarthmore Program. Swatties (our nickname for ourselves) tend to apply themselves to many things and go all-out for what they are interested in."

Students also pick up on and reflect the college's mission to teach more than just academic subject matter. "Swarthmore is special because of its close-knit nature. Many students feel a sense of community among about 1,350 other students. Students are frequently reminded to strive for 'ethical intelligence' and moral responsibility." Another student says, "One of the most important things to me at Swarthmore is the code of conduct that all students abide by. It's not a written code that every student signs like it is at some colleges, but more a way that everyone lives life at Swarthmore. It fits well into my Quaker heritage and that of the school (though it is nonsectarian now). I also really like the personal attention that students receive at Swarthmore. Especially after having spent a semester abroad at a large university (48,000 students), I appreciate that my professors know me by my first name instead of by a number and that they are so accessible."

Students at Swarthmore seem to defy generalization, as a junior points out: "There is no one type of student that does well at Swarthmore, and I can think of counterexamples for each adjective I am about to give you. Nevertheless, I would say that the majority of Swatties are intelligent, involved, self-motivated, hardworking, and need to know how to budget time. Being organized is a big help, because there is so much going on outside classes—you have to know how to get involved and keep being involved without drowning in all that you do. It's also important (and something that I would say most students learn pretty quickly at Swat) to search for what you want at Swarthmore, and if you can't find it, ask, or start it up yourself. There are a huge number of college-funded activities and organizations (I believe the number is over a hundred), and you'd be hard-pressed to find a student who isn't a member of at least two or three." Another student says, "People who are internally driven to learn do best at Swarthmore. A competitive atmosphere, which some people use to propel themselves through high school and other colleges, is nonexistent here, so personal motivation is a requirement. Additionally, the ability to write well and to know when to ask for help if needed (especially from professors, who are universally approachable) are very helpful skills to develop early. Quickly, after getting over the feeling of being less intelligent, socially aware, and politically active than everyone else, students realize they're 'all in this together' and really support one another. Resident assistants, peer academic advisors, and upperclassmen are also instrumental in helping students to succeed at Swarthmore through their constant availability to dispense advice, wisdom, and support."

Students enjoy the diversity of the college and the financial aid budget that helps to fos-ter it. Says one recipient, "The school's policy of need-blind admission has been particularly important to me and I am happy that Swarthmore is doing all it can to maintain that policy. The pursuance of this policy in the admissions process has made the college arguably the most diverse small liberal arts college in the country." One student also comments on the travails of small-college life: "One thing that I don't like about being at a small school is how fast gossip spreads. The entire campus is active, and a small incident can spread into a big issue before any authority has time to get the real story out to the students. It's part of what comes with being at a small school, though—and I would say that the benefits more than outweigh the negatives." Some students would like Swarthmore to enjoy more of the national and international reputation they feel it deserves.

For a recent graduate reflecting on her experience, "The school is excellent at providing a rich and diverse array of activities and speakers outside the classroom through a constant influx of scholars, professors, and social and political figures from around the world . . . The majority of students at Swarthmore (certainly by the time they leave) feel compelled to get out and change the world for the better, whether through the more typically thought-of avenues of medicine and law or through social justice work, innovative research, or grassroots organizing . . . I have absolutely loved my time at Swarthmore. I can't imagine a better place for my personal and intellectual growth on so many levels. I feel I have *definitely* come out with an increased social awareness, armed with the tools to challenge and evaluate the opinions and issues around me and to elicit positive, constructive change."

What Happens After College

Thanks to the caliber of teaching and the skills developed, Swarthmore graduates have an outstanding record of acceptance into graduate schools, with a particularly impressive record in the sciences and humanities. Swarthmore ranks among the very top colleges in the percentage of college graduates who attain the doctoral degree. Many enroll in medical school, engineering, education, business, and law. Their success in winning prestigious fellowships is impressive: In the class of 1998, five were chosen for Fulbrights; there were four Watsons and eleven National Science Foundation fellowships, plus some thirty other graduate scholarships. As the college states it, referring to actual alumni: "Swarthmore graduates edit national magazines, run worldwide advertising agencies, write *New York Times* bestsellers, start successful software companies, build manufacturing empires, teach schoolchildren, raise families, and run for president. Among other things." And they leave campus having established a very particular bond with their classmates—past, present, and future. As one administrator puts it, "Alumni administrators at our peer colleges often comment to me about the apparently unique and almost mystical bond that our alums seem to have for each other, regardless of age, geography, or politics. Our new vice president has observed that at social gatherings off-campus, our alums seem more interested in probing college policies and principles than in swapping gossip with each other.

He's also noticed that alums are more likely to tell him about the ethical values they learned or reinforced here than courses they took or the big game they played in."

VANDERBILT UNIVERSITY

2305 West End
Nashville, TN 37203-1700

(615) 322-2561

www.vanderbilt.edu

Number of undergraduates: 5,800
Total number of students: 10,200
Ratio of male to female: 51/49
Tuition and fees for 1999–2000: $23,598
% of students receiving financial aid: 55
% of students graduating within five years: 81
Endowment: $1,539,242,000
Endowment per student: $150,906

Overall Features

Vanderbilt is named for its founder, Commodore Cornelius Vanderbilt, who had in mind two goals in donating $1 million in 1873. The first was to create a university dedicated to the highest level of teaching and research that would benefit society like the top northern institutions, and the second was to contribute to strengthening the ties that should exist between all regional sections of our common country. Both were noble purposes as the nation worked toward healing its wounds following the Civil War and began its economic and technological expansion. A visit to the campus and a survey of the academic programs, teaching and research facilities, and graduate schools make it clear that this early vision has been fulfilled beyond what the generous commodore could have imagined. With a traditional undergraduate college of arts and sciences at the heart of the university and a number of specialized schools, Vanderbilt is similar to its Ivy League cousins. It is more diverse than some of them, however, in that there are ten defined academic schools that provide thousands of courses at the undergraduate and graduate levels.

There are 1,800 full-time faculty who are known for their commitment to teaching and advising students at the same time that they are expected to pursue research in their respective disciplines. Students come to Vanderbilt from all fifty states and ninety foreign countries. Of the student body, 44 percent are from the Southern states and the rest of the American students are from every corner of the country, with very large mid-Atlantic and Midwest contingents. Vanderbilt is situated on a beautiful campus of 300 acres on the edge of Nashville, a vibrant and growing city of 1 million residents. The Peabody College section of the campus is registered as a National Historic Landmark, having been designed after Thomas Jefferson's model for the University of Virginia. The main campus was designated a national arboretum in 1988. Thus students have the opportunity of living in a handsome, well-appointed environment with a city famous for its music, entertainment, and film industries, and good restaurants, performing arts centers, and shopping. Nashville is also the state capital and home to a dozen other colleges. While campus life is very active with arts events, athletics, student organizations, and its Greek system, students like the fact of having so many attractions available down the road as well.

Students choose to attend Vanderbilt because they want a balance of very strong academics, an active social life, a spirited community, a moderate political tone, an extensive range of curricula and majors or specialties, and excellent facilities. On the whole they are from cosmopolitan backgrounds and find the size of the university and the location very appealing. A majority are thinking ahead to graduate school and careers and believe that Vanderbilt's education and reputation will prepare them well. Most undergraduates—85 percent—choose to live on campus, which is a requirement for all first-year students. Approximately 35 percent of men and 50 percent of women join fraternities and sororities, which are one of the active social outlets on campus. With the many school-sponsored activities and outlets in the city, there is not a great deal of pressure to join a Greek house.

Vanderbilt commits a good portion of its large endowment to generous scholarships, physical facilities, and very competitive faculty salaries. It recently received the single largest gift ever given to a university from one individual, over $340 million for academic and financial aid programs. Over one-half of undergraduates receive financial aid in order to ensure that the meritocratic ideal of the opportunity for talented young men and women of all racial and socioeconomic backgrounds to attend the university is fulfilled. There is a definite commitment at the trustee and administrative level to diversifying the student body. Vanderbilt also awards 850 merit scholarships each year and 240 athletic scholarships to students of exceptional ability. Ninety honor scholarships are awarded each year to outstanding students of color, those with musical talent, leadership, community service or teaching commitments, interest in engineering and technology, or demonstrated exceptional intellectual ability.

Students are quick to compliment the university on the quality and range of facilities available for residential, social, physical, and academic life. The student recreation center is one of the largest and most comprehensive in the country. It includes a full-size gym, squash and racquetball courts, pool, weight-training center, a cafe and social activities room, and an 8,000-square-foot wellness center with full-time trainers. This facility, in combination with acres of playing fields, enables students to participate in any of forty different intramural teams and twenty-three registered club-level sports. The Starratt Student Center has just added a major theater and film wing, restaurant, and bistro. Under the leadership of its retiring chancellor of the past twenty years, Vanderbilt has refurbished virtually every residence hall and classroom on campus, in addition to constructing a number of new buildings for teaching and research. Vanderbilt is a member of the NCAA sponsored Southeastern Athletic Conference, which means they field a large number of varsity men's and women's teams that add to the campus spirit. However, due to Vanderbilt's high academic standards the teams find the competition from the large state universities in the conference a major challenge, particularly in football.

What the College Stands For

Vanderbilt inculcates a spirit of community caring through the honor code that has long been in place. Students are governed in their academic behavior by the principles of the honor code and tenets of community values.

Vanderbilt publishes each year a statement called the Community Creed, which establishes the principles of behavior and values by which the community is guided. They include the tenets of scholarship, honesty, civility, accountability, caring, discovery, and celebration. These are the values the university pledges to foster for the well-being of the entire community. In the spirit of its founding, the university is committed to providing students with excellent teaching and preparation for assuming a responsible role as professional and community leaders. The honor code and community service programs, the residential life, and extracurricular outlets are all designed to build individual skills, cooperation, and concern for others. The theme that learning takes place through a variety of activities in a variety of settings is strongly articulated. The individual must become self-sufficient and honorable in his or her dealings with peers and instructors. Volunteer service to

the community has long been a value that the university encourages; to this end, an Office of Volunteer Activities coordinates the twenty student service groups who interface with sixty community agencies and fifty local schools in myriad support roles. The Ingram Scholarship Program, established to honor the late chairman of the board of trustees, underwrites community service projects that students create and provides tuition scholarships on a selective basis to those who commit a good deal of their time to assisting others.

A senior administrator notes the educational goals of Vanderbilt: "a) to provide a strong liberal arts education which challenges students to develop their intellectual potential; b) to provide depth of learning in one or more majors as preparation for subsequent graduate and professional work; c) to prepare students generally for the 'offices of life' by giving them ample opportunity to learn how to work effectively with others, to lead and to follow; d) to instill in them an ability to reason, morally and intellectually, to recognize and develop their strengths in solving problems, to learn to apply what they have learned to the task of recognizing what needs doing and the ability to do it." Vanderbilt focuses a great deal of energy on its undergraduate students, seeking through a wide variety of means to help them succeed. These include "a high level of expectation from faculty and peers, but in a friendly, open, and nurturing environment; a strong academic advising and residential counseling system; the availability of faculty and staff who make getting to know and serving individual students a priority in all aspects of their work responsibilities; a widespread sense of civility in the community as a whole, combined with expectations of high industry from all members of that community; and pleasant physical surroundings—a beautiful campus, comfortable residence halls, student recreation center, etc."

What distinguishes Vanderbilt among many of the Hidden Ivies—and colleges and universities in general—is its combination of strong undergraduate teaching and education with a research institution offering significant resources and programs. As an administrator describes it, Vanderbilt has a "commitment to the finest undergraduate experience available which is enriched by being embedded in a leading research university, surrounded by nationally ranked professional and graduate schools; a commitment to supporting undergraduates with the kind of attention that gives students the space to learn to make their own choices, and the assistance and services to make those choices well informed; and an inordinate commitment to offer undergraduates the right balance of challenges and support required to prepare them for active, productive lives."

Curriculum, Academic Life, and Unique Programs of Study

Vanderbilt provides a sweeping range of courses and concentrations. The undergraduate college consists of four distinct schools: the College of Arts and Sciences, the Blair School of Music, the School of Engineering, and Peabody College of Education and Human Development. Students are required to experience a range of liberal arts and science subjects regardless of which school they are enrolled in. This is in keeping with the over-

arching mission of graduating thoughtful, humanistically educated students who have the perspective and judgment, the critical thinking, writing, and communication skills necessary to cope with the complex and rapid changes they will confront. The academic year is divided into two semesters and an optional May term in which one course can be completed for credit in any department. First-year students must choose two interdisciplinary seminars that cover a variety of topics. Each of the schools has its own set of core requirements that builds on the general curriculum.

The 3,000 students in the College of Arts and Sciences have access to twenty-five different majors, eleven interdisciplinary concentrations, and the opportunity to design their own major. Peabody College specializes in education at all levels and human development. In addition to centers for teaching and research in the traditional area of education, a center for entrepreneurship studies is one of the many special programs Peabody supports. Peabody's 1,000 students can major in dozens of fields that lead to careers in administration and teaching in schools or colleges, business, and human service agencies. Peabody is considered one of the very top education schools in the nation at both the undergraduate and graduate level. The School of Engineering has an enrollment of 1,200 undergraduates and offers eight specialized majors. The faculty and facilities are rated among the strongest in their fields. The Blair School of Music is one of the jewels in the university's academic crown, with over seventy faculty, many of whom are preeminent in their specialty, who instruct only 168 students, all of them undergraduates. The physical facilities are exceptional in all respects: a superb library of relevant books and musical scores, dozens of practice rooms, and a performance auditorium. Blair has received significant gifts enabling it to triple the size of the facilities and secure state-of-the-art equipment. Blair is committed to training only undergraduates, who thus receive the benefit of intensive training and advisement thanks to a faculty-to-student ratio of two to one. Blair musicians can major in music performance, composition and theory, musical arts, or musical arts with education. All majors are encouraged to elect courses from the other divisions of the university to round out their education.

To carry on its mission of educating students across so wide a spectrum of disciplines, Vanderbilt has created nine libraries, holding over 2.5 million volumes. The computer and technology resources are also among the very best to be found in an independent university. As a university with great financial resources and preeminent graduate schools, Vanderbilt can sustain many specialized centers, ranging from literary studies and creative writing to a program in moral leadership for the professions to the largest archive and study center of television news to numerous technical and scientific research institutes. One-third of all students will participate in one of the campus-abroad programs Vanderbilt sponsors in Spain, Germany, France, England, Italy, and Jerusalem. Thanks to its professional graduate schools, ambitious students can combine bachelor and master degree programs in five years of study in music/education, A.B./M.B.A., and B.S.E./M.S.E., among other 3/2 programs. The odds are small, indeed, that a student will not find a subject or field of concentration available to him.

College officials also note as important programs and policies: the student-run honor code, an "alternate spring break" where students participate in a variety of community service projects, the faculty pre-major and major advising system, residential life and education activities, and an office to assist students with disabilities. A college official, in identifying the importance of the honor code, also highlights some of the differences among the Hidden Ivies in how such codes or principles are defined and implemented: "The honor code is integral to the functioning of the university. It requires students to conduct themselves with the highest standards of honesty and integrity. Each student must sign the pledge certifying that he/she has not received nor given aid for every paper and test. In general, the honor code functions as a cornerstone of our residential community. Violations are handled by a student-elected Honor Council."

Major Admissions Criteria

In its search for students Vanderbilt focuses on academic preparation, involvement, and consistently strong performance and leadership. One administrator says that the university is looking for "students with proven records of academic and personal success in high school as evidenced by doing consistently well in a strong, college-preparation curriculum, and students who, by their extracurricular activities, indicate breadth and variety of interests, and abilities. Students do well who respond to a high level of faculty-student relations and who thrive in an environment of supportive, but competitive, peers of similar preparation, interests, and abilities."

Key admissions criteria are "a proven record of academic success in a challenging high school curriculum; consistency of performance at the highest level, academically and in extracurricular activities, or an upward trend in performance indicating growing ability and maturity; and an indication of interest in and special abilities to make a difference." Another administrator says, "Vanderbilt seeks students of academic curiosity, leadership, and personal character. The students who do well here are involved in many things beyond the classroom. All students study at Vanderbilt, but the most accomplished ones also are involved in community service, sports, student government. It's an active, friendly, all-encompassing campus." Important criteria for this admissions office representative are: "First, academic preparation. Second, level of involvement outside the classroom that speaks to a student's passion, leadership ability, and scholarly pursuits. Third (or really, tied with second) is teacher and counselor recommendations."

The Ideal Student

Thanks to its size, its great array of academic disciplines and extracurricular programs and opportunities, and its location, Vanderbilt can meet the needs and interests of most students. It holds special appeal for the smart and academically motivated individual who is determined to experience a blend of social, academic, and nonacademic activities.

The notion that Vanderbilt is "very Southern," with all the meanings contained therein, is simply not true. It does a great job of casting a wide geographic net to ensure a good mixture of students in the community. There is a feel to the environment that differs from that of northern tier colleges. Students define this as the friendly attitude of the students, faculty, and administrators, an interest in taking time to develop close social relationships, and a more traditional view of issues and events. The majority of undergraduates are outgoing, energetic, and committed to doing well in their studies while balancing them with an active extracurricular life. The workload is heavy and the grading competitive, but there is not a sense of cutthroat competition associated with many other mid-size teaching/research universities. Students consider their personal values traditional to conservative, which in part translates into a respect for family, concern for the disadvantaged and for threats to the larger community. There are few radical or highly politicized students and not too many such faculty visible on campus. There is lots of room for many kinds of personalities in the community, but the majority of students will be working that balance between the academic demands and their other interests.

In terms of social life on campus, fraternities and sororities are a major, though decreasingly important, force, and Vanderbilt and Nashville offer the interested student many alternative sources of enjoyment and involvement. As one administrator notes, "The university community is unusually friendly and supportive. Undergraduate life is rich in social activities, many of which involve a large sorority and fraternity base. And while the numbers of students directly involved with Greek organizations is decreasing, their influence on the social patterns of undergraduates remains high. A new Dean of Students Office is charged with increasing the opportunities for all students, regardless of their social affiliations. The commitment is extremely high at all levels of the university community to focus on the entire student experience and on making that experience more inviting for a wider range of students from many different backgrounds."

Student Perspectives on Their Experience

Vanderbilt appeals to students because of its friendly personality, its multitude of academic programs and opportunities, faculty involvement, and location. One first-year student describes "the welcoming feeling I got after being on the campus for fifteen minutes. The beautiful campus and friendly kids who went out of their way to make me feel at home for absolutely no reason other than the fact that that is the norm at Vanderbilt . . . The overall attitude of the faculty, staff, and students is warm, friendly, and relaxed, which is important." A junior says, "I think students at Vanderbilt, unlike at many schools of its academic caliber, manage a perfect balance of academics and social life. We know when to work, but we also know when to stop and have fun. Directly related to that, I think, is the relaxed academic atmosphere. There's none of the cutthroat competitiveness among the student body that seems to characterize other schools. If I ask a classmate to help me understand a given subject, he won't give me the wrong answer because we both might be competing for the same spot in a med school. If I do badly on a test, I won't be made to feel like I need to commit suicide. Also, no one's interested in how well you might have done in a class. Everyone is expected to do well, but if you talk about how well you're doing, no one's going to like you very much."

Some students note that their classmates can find almost any kind of student to befriend on campus. Others state that "students are almost clones of each other appearance-wise: 'Vandy girls,' preppy, blond, stereotypical 'rich' kids." And one student wishes the university would do more to "celebrate diversity, approach controversial issues, and become more supportive of minority programming . . . Our campus lacks socioeconomic diversity, but that is to be expected. Great strides have been made to recruit minorities, but little has been done to retain them. Vanderbilt has an atmosphere which is friendly and receptive on the surface. What makes it special is that it draws people from all over the world. The different cultures are available for us to explore, we just have to take advantage." Other students say, "There is a good, not a great, amount of diversity on campus. More should be done to draw in students of different backgrounds. Geographically, however, Vanderbilt has it down perfectly. Although the school is located in the South, the majority of students are from elsewhere." And fraternities and sororities still hold sway too strongly for some students. "Greek life, while by no means dominant, is a major part of the social scene on campus. Sure that can be fun, but it gets old quickly. The Greek scene, I think, should somehow be restrained so that it isn't the only option. Of all the people I know who've left Vanderbilt, all but one left because of not wanting to join Greek organizations but feeling enormous pressure to do so."

Those who do best at Vanderbilt are

"focused, goal-oriented, down-to-earth kids. The 'grade grubbers' do not make it here. Kids who are balanced academically, socially, and individually." A junior says, "I think generally every group does equally well. Obviously the student who works nonstop will do better academically than the one who parties every night, who will have more fun. The great majority of students, I feel, has learned to balance these two effectively. I think it is these people who benefit most from the college experience—those who work when they need to work and who are social when they need a break."

What helps students to succeed at Vanderbilt is, according to one, "The willingness of the community overall to help anyone who asks for it—academically, emotionally. Every professor I have encountered has been more than willing to help in any way they can." What makes the school special is, "It is a top-twenty university minus the overly competitive, stressful, snobbish environment. I think the best thing about the school is that I feel I can approach virtually anyone on the campus and get a warm, responsive reaction. I am not intimidated to limit myself to my own clique or group." Another student says that "a relaxed academic atmosphere fosters a good work ethic. Without the constant pressure that other schools seem to pride themselves on, I feel more compelled to work hard and succeed." A senior says, "The abundant resources and the availability of the administration, and an atmosphere conducive to learning" help students to do well.

"I have been astounded at the number of programs and groups that Vanderbilt offers," says a freshman. "Students can find anything they want." A junior adds, "Vanderbilt really is unique in that it has an excellent academic reputation and Division I, SEC sports. On this level, I would put it on par with more recognized schools like Northwestern and Duke. However, Vanderbilt also has a singular easygoing, non-pressured atmosphere—something most schools of high academic caliber cannot say. This is what makes the school stand out. Also, its student body is relatively small. I can say that there are days I can be walking around campus and see no one I know, and there are days I can walk around and see everyone I know. It's large enough not to be confining and boring, but small enough not to be impersonal. Also, for me personally, Vanderbilt's Southern location adds something important to the school. Being from New York, I'm used to people being unfriendly and largely uncaring. Then I came to Nashville. The first time someone I didn't know said 'hi' to me, I thought they were either going to mug me or had some kind of mental imbalance. In that respect (hospitality and friendliness), the school is very Southern."

What Happens After College

The majority of Vanderbilt graduates will eventually obtain advanced degrees, with some 70 percent attending professional and graduate schools within five years. About 63 percent of students go directly into the workforce and 35 percent go directly to graduate school. Many Vanderbilt graduates find work immediately after graduation in national and international businesses. One administrator says, "Graduates are prepared and equipped to continue to learn—they know how to accept challenges, overcome obstacles, and succeed."

VASSAR COLLEGE

124 Raymond Avenue
Poughkeepsie, NY 12604-0010

(914) 437-7300

www.vassar.edu

Number of undergraduates: 2,360
Total number of students: 2,360
Ratio of male to female: 40/60
Tuition and fees for 1999–2000: $24,030
% of students receiving financial aid: 60
% of students graduating within five years: 83
Endowment: $555,000,000
Endowment per student: $235,170

Overall Features

Vassar was founded in 1861 as a traditional residential liberal arts college by Matthew Vassar, who apparently wanted his name associated more with the education of women who had the capacity to become independent, spirited leaders than with the brewery products that earned him his fortune. Like women at its sister institutions founded in the same period—Smith, Wellesley, Mount Holyoke, and Bryn Mawr—Vassar women were encouraged to think for themselves and presume that they were the equals of the young men of their station being educated at the traditional Ivy League institutions. The graduates of these early women's colleges have had a significant impact in the civic, artistic, educational, and professional spheres from the earliest years to the present. Vassar first admitted men in 1968 in recognition of the changes occurring throughout higher education at the time, including the coeducational movements at most of their peer institutions. Given its high selectivity of students through its first century, the college was determined to maintain its ability to admit intelligent students who demonstrated a true passion for learning, questioning, and taking a role in vital contemporary social issues once they graduated.

Today's student body reflects these early goals of the college. Vassar continues to offer a rigorous curriculum taught by a distinguished faculty who expect their students to think for themselves, analyze critically the topics and issues at hand, and demonstrate strong communication skills in their writing and speaking. The great majority of classes are small and require plenty of student participation. Vassar classes average seventeen students, and the faculty-to-student ratio is ten to one. One advantage Vassar students enjoy over students of most colleges is that 80 percent of the 200 faculty members live on campus, so their involvement with students occurs as much out of the classroom as in it. Both the political and academic tone is definitely liberal, open-ended, questioning, and free-spirited. If there is any conformity to be found it tends toward acceptance of the concerns that spark campus debate, and activist causes typically have to do with gender issues, racism, politics, and sexism.

Matthew Vassar and the early leaders of the institution would no doubt be surprised to witness the changes in the viewpoints of the student body. The earlier women admitted to Vassar were predominantly from families of high social standing and wealth, mainly from the Northeast, who developed over time an attitude of noblesse oblige with regard to their role in society. Contemporary students clearly are caught up in the many concerns our society is trying to cope with. They are, by contrast, a highly diverse community of talented men and women in keeping with the college's celebration of the diversity of people, ideas, and opinions. One-quarter of the students are of color and 60 percent of all students receive financial aid. While Vassar cannot guarantee aid to all qualified candidates, it does meet the need of the majority with generous aid packages. There are no academic merit scholarships, as the college commits its resources to those who demonstrate financial need. Today, 60 percent of enrolled students graduated from public high schools, which also represents a major change from the college's earlier constituency. Thanks to the passionate feelings a great many earlier alumnae had for the education and val-

ues they gained as undergraduates, their financial generosity has built a robust endowment and superb physical facilities.

Although it is located in an old and tired industrial city, Vassar has 1,000 acres of beautiful grounds with 100 buildings for students to enjoy. There are multiple residential living options, from the older, elegant suites to newer dormitories and small houses. Virtually all students live on campus for their four years. There are no fraternities or sororities, so social life revolves around the residences, student center, and the many student organizations.

The library has extensive holdings of books and manuscripts and is fully computerized for searching for materials, as well as for providing access to other sources via national and international networks. There are separate art and music library collections as well. The entire campus is networked for immediate access to libraries, faculty, labs, and the Internet. The concert hall, studio art center, and theater are first-rate.

The college is intent on building a stronger science program and facility as more students are interested in combining their arts and humanities training with the sciences. Student life is enhanced by the presence of a refurbished student center, a new $10-million, multiple-use athletic facility that is nearing completion, and a nine-hole golf course. The college sponsors twenty-four varsity athletic teams for men and women that compete at the NCAA Division III level. Many club and intramural organized teams are also available. As a result, three-quarters of the student body participates in some level of organized sports— rather impressive for a college thought of as a progressive, intellectually oriented community with little traditional school spirit. Vassar students also participate in a great many extracurricular programs, from a very competitive debate team to theater and music performance. Some students devote their energy to theme or affinity groups to make certain that their interests and needs are recognized by the community at large.

What the College Stands For

Vassar's educational leaders believe passionately in the positive role their graduates should play in the professions, the arts, and the community at large. Concern for the future of our planet, the rights of others, and a standard of ethics and humaneness in the professions and commerce are some of the themes that are articulated through daily life on campus and in the classroom. Leadership, intellectual independence, learning how to think and react critically, and the breaking down of artificial barriers are the major attributes Vassar hopes to inculcate in its students. The nature of teaching and the opportunities for different forms of learning as described below are intended to fulfill these goals. The overriding theme is the quality of the faculty, their dedication to teaching and guiding students, and their passion for and expertise in their discipline. "Nothing is more important to a college than the quality of the faculty," says a senior dean. "We recruit the best minds of each generation and we support them when they are here. Our support includes aid to conduct serious scholarship and a strong program to strengthen teaching skills." The administration and trustees are committed to upgrading particular facilities in order to maintain the

caliber of programs and learning opportunities beyond the classroom. Financial resources have been dedicated to upgrading the athletic facilities and expanding athletic activities, as well as refurbishing the theater building and classrooms. The development of information technology and the use of computers and the Internet have been other changes at the college.

A senior administrator identifies some of Vassar's educational goals: "Through careful planning that combines independent action on the part of the student with wise counseling on the part of faculty advisers, Vassar offers a wide range of educational opportunities. Our liberal arts curriculum stresses depth of study within a discipline while providing breadth of exposure to many areas of inquiry. This depth-and-breadth approach results in intensive involvement in scholarly work, whether it be scientific research in a faculty member's lab, the writing of poetry with a member of the English department, or conducting archival research with the guidance of an eminent historian. Whatever the path chosen by a Vassar student, he or she will graduate knowing how to write and think, and will be open to and ready for the challenges of graduate or professional study or the demands of the business world." A dean notes these educational goals for students: "With the help of their professors and advisers to develop and carry out a program of study that balances curricular breadth and depth, engages them with different modes of thinking, has intellectual coherence, and meets their own educational goals; to learn to question assumptions and received ideas; to take a position and support it with an argument (which also involves learning the distinction/relationship between opinion or personal response and argument); to articulate arguments clearly, both orally and in writing; and to acquire general skills in critical thinking and quantitative analysis that can be applied in a wide range of contexts after graduation. Ideally, a college education gives you not only a certain body of knowledge but also an understanding of how new knowledge is acquired: We hope to teach our students how to teach themselves."

The college sees its history and philosophy as contributing to its current unique identity: "The origin, growth, and transformation of Vassar from the mid-nineteenth century to the present have created a liberal arts college of unique distinction. Our commitment to the highest standards of intellectual discourse and scholarly achievement coupled with a social setting that provides respect and support for all, women and men, whatever their race or ethnic origin, make Vassar different from and—we truly believe—better than other coeducational institutions." A dean continues, "With its incredibly strong heritage as a pioneer in the education of young women (at a time when most didn't even go to high school), Vassar has always stood for equal opportunity . . . Underpinning this equality is the very fact of the institution's existence, and as we approach our thirtieth year of coeducation, we celebrate the fact that neither women nor men have had to 'move over' to a secondary place to allow for the other. I have heard our men and women graduates speak so eloquently of this equality and describe how it has prepared them for their professional and personal lives."

Curriculum, Academic Life, and Unique Programs of Study

There are no core requirements necessary to graduate other than two courses selected from a large menu, one that emphasizes oral and written expression and one focusing on quantitative analysis. Vassar encourages students to explore a broad range of intellectual fields and to choose a field of concentration that they will delve into with great zeal. Well over forty majors and thirty interdisciplinary concentrations (such as American culture, women's studies, and environmental sciences) are available, with over 1,000 courses to choose from. The admissions dean states that the array of options for students is truly remarkable. Conversations with various deans regarding the special features of the academic program revealed the common theme of students taking charge of their educational experience by virtue of the flexibility of the academic program. The key to realizing their goals is the commitment of the faculty to advising and supporting students, not just transferring information in a formal classroom setting. Faculty provide personal guidance to the many students who wish to design their own field of study.

Vassar is one of only a handful of major colleges where faculty members live in each of the nine student residences. They are committed to a three-year term so that they form strong relationships with their students. A positive reflection of the willingness to engage with undergraduates is the waiting list of professors who apply for house residencies. It is very easy to create an independent study program or an interdepartmental major or double major under the aegis of a professor. The primary method of education the faculty emphasizes is to teach students how to execute firsthand investigation using primary texts, analysis and synthesis of information, and on-site learning. The college states that "in a typical semester, seventy-five to one hundred students pursue independent study, in everything from political psychology to the physics of black holes. Five percent of each class creates independent majors, combining independent coursework with available offerings in various disciplines. One faculty member, a biologist, put it this way: 'Our students engage in research at a level they would not be able to approach at a research university. So far this year, I've submitted three papers for publication, all of them with student coauthors.'"

All students must demonstrate proficiency in a foreign language either through placement testing or completion of intermediate-level coursework. The college encourages students in all fields of study to spend a term or a semester on one of the multitude of approved foreign study programs that can find a Vassar student in Africa, the Middle East, the Far East, Australia and New Zealand, Russia, Eastern or Western Europe, or South America. College-sponsored opportunities abound for research and internships in the natural sciences, the performing arts, the humanities, and social sciences. Vassar is a member of the Twelve College Exchange program that lets students spend a year at any of the other colleges. Students confirm that they have a good deal of flexibility to develop a program of study that suits them and that the faculty is happy to assist them in formulating their ideas and structuring an appropriate set of courses and

individual study and research. Strength in writing and critical thinking are hallmarks of the skills the faculty wishes to develop in students irrespective of the field of study a student pursues.

Major Admissions Criteria

Vassar is very selective in the admissions process, although it is not as fiercely competitive as the Ivy universities and some of the other Hidden Ivies. The great majority of applicants are bright, high-level performers who have succeeded in a demanding curriculum and have a variety of talents and interests. The dean of admissions describes Vassar students as being feisty, meaning independent in their attitudes and approach to learning. They are willing to take a personal stand on ideas presented to them. Thus, in addition to a challenging high school course of study covering all of the traditional subjects and strong grades, the admissions committee gives considerable weight to those who demonstrate clear evidence of a joy for learning; stretch their capabilities by taking the highest-level courses; and display independent thinking and leadership through clubs, sports, or student government. Relatively less weight is given to standardized testing as the committee believes test results do not reveal the personal qualities that predict success and add value to the Vassar community. Another administrator identifies "independence of thought, accomplishment in a range of the most challenging high school courses, and personal integrity" as the most important admissions criteria. How does the committee ascertain these traits? Through the courses a student has taken in high school; teacher and counselor recommendations; reports from employers, coaches, and advisors of clubs; interviews with alumni; and the applicants' personal statements in the application. We can attest to the personal attention and care the admissions committee gives to every applicant and their appreciation for the unique applicant.

The Ideal Student

Conventional thinkers and those in need of a highly structured academic format will not appreciate the more individualistic nature of a Vassar education. Nor will those students who flourish in a socially conventional or a highly competitive athletic community. The Vassar environment calls for an open mind on intellectual and social issues. Active learners are the most successful students. As one administrator puts it, "Open, inquiring, serious, and engaged (and engaging) young men and women flourish here . . . High intelligence, therefore, is assumed, as is a capacity for active involvement and participation in the classroom and in extracurricular activities, including, increasingly, athletics. Vassar students are not passive learners, regurgitators of their instructors' lectures. Rather, they are taught to think for themselves and to express their thoughts clearly in writing. Much reading and writing are expected of our students, whatever the curricular area, and the students themselves are the fortunate beneficiaries."

Another dean identifies these qualities as important: "Self-motivation. Vassar students must learn to take responsibility for their own education and to know for themselves why it is important to them. Initiative. Vassar offers its students a rich variety of opportunities and a great deal of freedom to set their own academic course, but it also expects them to speak up for themselves, to pursue their own interests, and to ask for help when they need it. A tolerance for freedom (and the conflicts and uncertainties that go with it). Students who need a highly structured program, who want decisions to be made for them, or who prefer to live in a homogeneous community will find Vassar a difficult place. And, a *passion for something.*"

Vassar graduates will enroll in graduate schools at a rate equal to that of students in other top colleges, but they will take advantage of their undergraduate years to experiment with an eclectic range of courses or concentrations and explore unusual opportunities off-campus. Grade competition and a tone of pre-professionalism are not in evidence in comparison to the more conventional colleges. Strong communication skills and an aptitude for independent learning are requisites for success. The individual who has little concern for the political agenda and social issues that flourish on the Vassar campus should look elsewhere.

Student Perspectives on Their Experience

Students choose and enjoy Vassar for its flexible educational program, academic strength, location (a beautiful campus two hours from New York City), and social climate. Vassar "gives each student the flexibility to explore their interests without putting restraints on them academically or socially," says a senior. "I like that Vassar pushes me to do more and always want more." Another sees as most important "the atmosphere. No two people are alike on campus. Since I came from a very 'cliquey' high school that was all about fitting in with the status quo, Vassar gave me a chance to be myself. I feel as if I'm liked for who I am, and not where I came from, who my friends are, what clothes I wear, etc. I felt that atmosphere at Vassar during my first visit here. It definitely wasn't like the other colleges I had visited. The other colleges reminded me too much of high school." "For me," says another student, "I thought the best undergraduate education I could receive was at a small liberal arts school. Vassar has an excellent reputation among such schools and I knew I could play soccer here as well. The campus is gorgeous, the students are friendly, and I enjoy the suburban environment so close to New York City."

Students have difficulty identifying a type of student who does best at Vassar, but generally feel that independent students who like a noncompetitive atmosphere will get the most out of their experience. "I think everyone excels each in their own way and in different things," a senior says. "The great thing is that there is no pressure to be number one in your class, like in high school. It's more important to do whatever it is you do to the best of your ability and not worry so much about everyone else . . . You really have to be a strong individual in this strong environment, because while Vassar is a great place to be, it is very challenging academically, and you really have to be ready." Another senior says students do well who are "outgoing and active individuals who make it a point to participate in extracurricular activities and/or nourish their nonacademic interests."

Vassar students succeed because of "the friendships you make," according to one student. Another says, "I feel as if it's the professors' desires to 'want' their students to excel. They make an effort to get to know their students and make themselves extremely approachable. They encourage questions and feedback and make an effort to tailor their courses to our wants and needs." And "Vassar provides an extensive advising system. Each freshman is assigned a Student Fellow who can advise them informally on personal and academic matters. Also each is assigned a pre-major adviser, then later a major adviser and thesis advisor."

Such offerings as the Junior Year Abroad program and the college's generous financial aid budget are identified by students as important elements of their Vassar experience. A senior says, "Through Vassar, I was able to study abroad in Madrid. My professors there were visiting Spanish professors from the local universities. I had the privilege of taking a course with a European icon theater critic. Every class was exciting because he spoke of actual experiences with working with actual contemporary artists. He knew and was known by everyone. It was an experience that I was able to have through Vassar, and for that I am grateful." Another senior notes, "The fact that there are over a hundred student groups on a campus of twenty-four hundred is amazing! There are so many options to be in whatever you want to be in, no matter what your major, your interests. It's so easy to try something new and there are so many things to keep you busy (maybe a little too busy at times!)." Vassar is seen as a diverse campus in many ways, with a supportive environment overall. As one student quips, "We were diverse before diversity was cool."

Students leave Vassar challenged and changed. "Vassar makes us think," says one. "It makes us start a dialogue with each other and encourages us to continue it wherever we go." Another senior says, "Vassar has done a fabulous job of helping me to think more critically, creatively, and constructively. I feel as if I am no longer looking at the world with rose-colored glasses. Instead of ignoring injustices, Vassar has taught me to confront them and to try and change them instead of going with the flow." Another adds, "The challenging classes prepare you to be responsible, attentive, well-read, and articulate." And for another senior, "Vassar is a residential college with a very open, socially aware community. The students here are independent, free-thinking, and often very talented. The academic environment is intense, but there is *no* competition between students for grades. Students here are also given a great deal of academic flexibility, since there is no core curriculum. The most important factor that dictates to what extent a student has a positive experience at college is the people he or she meets. Vassar excels on the people front."

What Happens After College

Vassar believes that a strong indicator of the quality of the intellectual skills and substantial value system students acquire is their success in a wide range of careers and the high rate of admission to graduate and professional schools; also, the many competitive fellowships (including Fulbright, Rhodes, Watson, Goldwater, and Marshall) and top national awards they garner each year. Over 80 percent of alumni will earn a graduate degree eventually. A senior college official noted, "We intend to remain rooted in the liberal arts tradition, because our education has proved to be the sturdy foundation for flexibility in professional growth after college. The philosophy major may become a physician, the biology major a journalist, the English major a banker. Making such possibilities available to our students will continue to have high priority." While a good many will train for the professional fields of law, business, medicine, and education, others will undertake studies in the various arts fields, social work, journalism and communications, and public service. Prospective Vassar students are told of the power of the alumni network that helps undergraduates and recent graduates to land worthwhile internships and full-time positions in competitive fields. There is much truth to this promise, as many Vassar grads will attest to.

The college reports that alumni identify a number of factors as being particularly important to their development: "Vassar's small size and the fact that virtually all students and many faculty live on campus, which allows the development of a thriving sense of community; the strong emphasis on individuality—extending even to the development of personal academic programs by students under the guidance of professors; the emphasis on equality between the sexes, that women and men learn to work and study side by side; the astounding variety of extracurricular offerings that give students opportunities to explore new interests and learn the skills of administering organizations and working in groups; and the rigorous academic requirements, which leave alumni feeling that anything is possible if they have earned their Vassar degree."

WAKE FOREST UNIVERSITY

P.O. Box 7305, Reynolds Station
Winston-Salem, NC 27109-7305

(336) 758-5201

www.wfu.edu

> Number of undergraduates: 3,900
> Total number of students: 6,000
> Ratio of male to female: 49/51
> Tuition and fees for 1999–2000: $21,420
> % of students receiving financial aid: 29
> % of students graduating within five years: 84
> Endowment: $748,000,000
> Endowment per student: $124,666

Overall Features

In 1834, the North Carolina Baptist Convention established Wake Forest to train young men in the region for the ministry or for lay leadership in the Southern Baptist Church. Wake Forest admitted women in 1942, and in 1956 the university relocated its campus from Wake Forest, North Carolina, to Winston-Salem. The Baptist sponsorship carried forward until recent years when tension between the goals and strict traditions of the Convention and those of academic freedom and intellectual inquiry came to a head. As the university grew in stature, thanks to the vision of the administrative leaders to place Wake Forest on the national map as a top small university, it attracted an increasingly more selective student body from outside the state and talented faculty who wanted the opportunity to carry on advanced research and teach a more diverse cohort of students. While teachers were frustrated by the limitations on the topics they could discuss and on their critical assessment of religious, philosophical, and scientific concepts, the students felt constrained by restrictions on drinking, dancing, and intimate dating that governed campus life.

After considerable discord and antagonism on both sides, the university dissolved its relationship with the Baptist Convention in 1986. The board and academic community voted to become an independent, non-sectarian institution. This is not to say that the ethical and moral teachings that had permeated the school for 150 years were also suddenly erased. Wake Forest continues to emphasize its traditional values that are based on the teachings of the Judeo-Christian tradition. Ethical behavior is infused throughout the community. Today, it is most pronounced in the emphasis assigned to the honor code, which governs the academic and social life of the student body. Wake Forest remains a conservative social and academic environment compared to the majority of colleges with which it compares itself today. However, it is highly successful in raising money from its loyal alumni to create professional advancement programs for its faculty and scholarships for talented students. It is also asserting its presence as a top-flight liberal arts college that cares greatly about excellence in teaching and educating young men and women in the critical learning skills. The aim of the undergraduate experience is to bring together accomplished academics who love teaching and care about their students. Their mission is to challenge students to think critically and extend themselves to their fullest capacity. From the flattering comments of many students, the faculty is succeeding both in stimulating and challenging them. There are 344 full-time teachers available to students across a wide spectrum of disciplines. Every leader of the institution has expressed his or her admiration for the outstanding quality of the faculty and their commitment to teaching.

The heart of the university is the college of arts and sciences, which represents the largest enrollment. There is one other undergraduate division, the School of Business and Accountancy, which enjoys an outstanding reputation for training students for the world of finance, accounting, and business administration. It has a proactive outreach program to attract minority students, scholar athletes, and students from other regions of the country. Until recently the percentage of students who receive financial support had not been as

high as in a majority of the other Hidden Ivies even though the university can operate on a need-blind admissions policy. The administration has recently allocated funds for sixty scholarships for minority students. At present 11 percent of students are of color and over one-half of all undergraduates are from the greater South. There is also a generous number of academic merit and athletic scholarships. Wake Forest has much to offer a bright student who wants to experience a traditional educational and social environment that puts a premium on good teaching and faculty interaction with students.

The physical campus is very attractive and well-equipped in all respects. The great majority of students who choose Wake Forest are attracted to the strong blend of academics, athletics, and social traditions that epitomize the school. Athletics are definitely a major component of student life. Wake Forest is a member of the Atlantic Athletic Conference, which is one of the most competitive in NCAA Division I-A sports. As the smallest member of the conference, Wake Forest has to attract a high percentage of student athletes. Its success in intercollegiate competition is exceptional when the university's size is taken into account. School spirit is a defining feature of campus life, as it revolves around athletic competition. The athletic facilities for men and women who compete in intercollegiate and intramural athletics are outstanding and a major selling point for the university. The other element that defines the social environment is the large Greek system, which is comprised of fifteen fraternities and sororities. Affiliation with a Greek house is high—40 percent of men and 47 percent of women. The administration is reevaluating the traditional role of the Greek presence and the advisability of first-year student rush, as are many other selective residential colleges.

What the College Stands For

Wake Forest is clear about its purpose as an educational institution and about what it means to be well-educated. "The undergraduate school of arts and sciences is the center of the academic life; through it, the university carries on the tradition of preparing men and women for personal enrichment, enlightened citizenship, and professional life. . . . It seeks to honor the ideals of liberal learning, which entail commitment to transferring cultural heritages; teaching the modes of learning in the basic disciplines of human knowledge; developing critical application of moral, aesthetic, and religious values; and advancing the frontiers of knowledge through in-depth study and research and applying and using knowledge in the service of humanity." This is a bold and ambitious mission statement that is implemented through a traditional arts and sciences curriculum that is informed by humanistic values and ethical issues and steered by a very committed faculty. Although the formal ties with the Baptist Church have been severed, the moral code of behavior and values is still alive. "We have adhered to high academic standards and a liberal arts philosophy enriched by an emphasis on personal responsibility and Pro Humanitate," says a senior dean. Indeed, the college asks all applicants to assess its Pro Humanitate (for the good of humanity) mission in an application essay.

The college sees student success as emanating in large part from student-faculty interaction and student support services and attention. Officials note the importance of "a careful admissions process, comprehensive orientation, advising, low student-teacher ratio and much faculty-student contact, and support services and resources"; "close relationships with faculty, a strong sense of community, leadership and training opportunities, the large number of clubs and organizations to become a part of, and great technology." "With its small class size and low student-to-faculty ration," the college argues, "Wake Forest ensures that there is time and opportunity for all points of view to be shared and explored, and that students receive the attention and intellectual stimulation for which they chose this small, private university. Teacher/scholars encourage critical thinking and analysis, encouraging students to form and express their thoughts and opinions."

In the future, Wake Forest will likely advance in its educational delivery and academic programs, but do not expect this institution to lose sight of its traditions. Says one dean, "I think Wake Forest will change in degree but not in kind. We will stay abreast of innovative initiatives in teaching and scholarship but not abandon our purpose and heritage." Another says, "Wake Forest will certainly change in the future; it is also a very stable environment where heritage, tradition, and universal values remain intact. At the same time, Wake Forest is cognizant of cultural changes that impact its people and programs." The university tries to combine the best features of a research university and a small liberal arts college and the important aspects of tradition with educational changes.

"Wake Forest has stayed faithful to its liberal arts curriculum and has continued to stress that students take courses in a planned and systematic way so they are exposed to the liberal arts in a sensible way. Wake Forest has also been innovative in regard to new initiatives, such as our use of technology to teach. We have remained small and flexible. We combine, I think, the best of the old and the new." Another dean adds, "Wake Forest consciously strives to combine the best features of a liberal arts college with the best features of a university. The dedication of our faculty to both research and close interaction with students is remarkable and highly unusual. The resources made available to students and the attention given to them are also unusual."

Curriculum, Academic Life, and Unique Programs of Study

To achieve its educational goals, the university requires students to gain exposure to subjects across all of the academic disciplines. The requirements are fairly specific. There are fifty majors to choose from in the liberal arts college that can meet the special interests of virtually all students. The student with unusual or very specialized interests who desires more flexibility than the structure curriculum provides can opt for the Open Curriculum Program, which lets him develop his own field of study with the close supervision of a faculty adviser. Another opportunity for the highly motivated student is the Research Fellowship Program, which is designed to match an undergraduate with a professor who is carrying out advanced research in his area

of study. In addition to providing invaluable experience in research and writing, the program also awards $2,000 financial scholarships to each research fellow. The university encourages exposure to other cultures through its study-abroad programs, including its own residential centers in London, Vienna, and Venice. Wake Forest faculty accompany students and direct a curriculum that for one semester focuses on the culture of the particular country. Over 35 percent of students participate in international study. Wake Forest has high expectations for its students and thus the academic load is heavy and challenging. Requirements to graduate are well-defined.

An administrator notes that "as part of a new plan to enhance undergraduate education, Wake Forest now provides IBM laptop computers and Lexmark color printers to its students. The plan, which contains about three dozen features, has also led to the hiring of more faculty, the introduction of specifically designed freshman seminars, new study-abroad scholarships, and the annual presentation of a series of events with a common theme." *Yahoo! Internet Life* ranked the university the third "most wired" campus in the nation in 1999. Wake Forest also encourages volunteerism and service, offering many opportunities for students to make a difference. "Rooted in its liberal arts tradition and 'Pro Humanitate' motto," says an administrator, "Wake Forest University strives to use knowledge for the benefit of humanity. In this spirit, students and faculty commit tens of thousands of hours to community service each year. Student volunteerism is reflected throughout clubs and organizations, including the Volunteer Service Corps, which serves as a clearinghouse for more than forty-five agency and school volunteer programs. In addition, special programs include volunteer service trips to Calcutta, India, and the Agalta Valley, Honduras, as well as seven domestic sites. More than 1,200 students volunteer weekly in service sites and up to 2,200 participate in service projects during the year. In addition to volunteer programs, the university also participates in the nationally recognized America Reads tutoring program, which involves federal work-study students in four elementary school sites providing academic help and mentoring. Our faculty are also committed to these service ideals. Through the Academic and Community Engagement Fellowship Program, faculty integrate service-learning opportunities into their existing courses. As a result of this program, students are able to synthesize their learning experiences both in and outside the classroom while also providing valuable service to the greater community."

Major Admissions Criteria

Wake Forest seeks "high achievers, academically curious, motivated students with strong character, students who value community and service to others," says an admissions officer. Top admissions criteria are "academic achievement, character, and motivation." A senior administrator says, "We look for academically bright students who are well-rounded and have a sense of purpose. These types of students do well at Wake Forest." And, he stresses, "Wake Forest looks at each application individually." For 1998/99, Wake Forest had 6,838 applicants for freshman admission, of whom 2,848 were accepted. Of these 965 enrolled.

The Ideal Student

Wake Forest is a distinctive institution in terms of its traditional social and intellectual culture and academic structure. Those who want to live with a group that is enthusiastic about athletics, a predominantly Greek social system, and a demanding, traditional approach to learning will find Wake Forest to their liking. Students should prepare for Wake by taking a broad and challenging high school course load, and should be interested in and excited about continuing to study a wide range of topics in college. The rewards of the Wake Forest experience are considerable: the accessibility of the faculty and the quality of its teaching, the camaraderie accruing from the residential life programs and active Greek system, and the opportunity to participate in or cheer for top-flight athletic teams for both genders. Do not consider Wake Forest if your priorities differ dramatically from these. Also, one will find only some, but not many, "alternative" students (students who are very far out of the social mainstream) at Wake.

Student Perspectives on Their Experience

Students are attracted to Wake Forest for its location and beautiful campus, small size and faculty attention, school spirit, and traditional values. "Wake Forest offered an incredible liberal arts education in a great environment," says a senior. "I knew the smaller classes would be beneficial and I loved the location. Wake Forest was one of the most beautiful campuses I had ever seen, and the care they gave to the grounds demonstrated the attention Wake seemed to give to every aspect of college life. Additionally, I liked the idea that my classes would be taught by professors, not teaching assistants and grad students. From the very start of my application process, Wake Forest gave me individual attention, and I knew that would continue throughout my college career."

Another senior reflects on his first visit to campus: "I can still remember my September visit to Wake Forest as part of the North Carolina Challenge Program, which invites students from around the state to spend the night at Wake Forest, to tour its facilities, attend classes, and meet many of its professors. I will always remember attending an eleven A.M. religion class with Dr. Ralph Wood. He lectured on decisions and their effects on our lives, using one of C. S. Lewis's novels as his text. He described God's role in our decisions, and as I looked around the classroom, I noticed quickly that each student was just as intrigued as I was. The class piqued my interest, and I was disappointed only when the fifty minutes were at an end. I was so impressed by the class that I knew that Wake would be among my top choices for college. Yet what happened that afternoon made it my only choice. After the closing session of the NC Challenge weekend, I was walking on the main quadrangle and I happened to pass Dr. Wood. I stopped him and started to reintroduce myself, but he cut me off. 'Tom [name changed here], how was your time on campus? I do hope you come here.' I was so surprised that he remembered my name, all I could do was respond with something like 'It was great' before thanking him and walking on. I thought a lot about that late-afternoon

conversation as I made my college choice. I decided I wanted a place that felt like home, and at Wake Forest, I had found that place. I am still amazed at the familial atmosphere to be found at the university."

What Wake Forest does best, according to students, is foster connections between students and their faculty and peers. Says a senior, "I think Wake Forest has an incredible skill for getting to know each student and discovering his or her gifts. Rarely will you find a school at which you can walk around campus and see a half dozen faculty or staff members who have touched your life in such a way. Every one of my professors has gotten to know me as an individual, has discovered the things I have to offer in class, and has helped me to grow as a person as a result of the class. My professors have supported me during tough times and I have been taught incredibly valuable lessons as well." Another senior says, "I would say that Wake Forest does three things incredibly well: First, it emphasizes technological competence to its students; second, it emphasizes collegiate community; and third, it provides opportunities for its students that expand their horizons and challenge them to be better students and better people."

Diversity could use some work at Wake, where students feel the environment is friendly, but not as culturally mixed as it could be. Says one, "I think diversity is a problem at Wake Forest, although efforts are being made to remedy this problem. After coming from an incredibly diverse high school in South Florida, in which students spoke myriad languages and shared their diverse cultural backgrounds, it was off-putting to see the lack of differences at WFU." Another critique centers on the university's "over-regulation" of student life, including a recent controversial move to control students living off-campus. Nevertheless, the university is reported to be fair in its dealings with students and open to and encouraging of student involvement in and discussion to resolve issues. Finally, students should be aware of the very strong role the Greek system plays on campus, for "while Wake Forest is officially 50 percent Greek to 50 percent independent, on the weekend it frequently feels more like 70 percent Greek to 30 percent independent students." And as a student notes, "Though I definitely wouldn't describe Wake as segregated or racist, we do need to work on intermingling to appreciate the diverse cultural and religious backgrounds represented. People tend to be politically correct; one rarely sees a student with crazy-colored hair or extremely alternatively dressed. The very small lesbian and gay population is supported by 'straight' students in the Gay-Straight Student Alliance." Social life outside the Greeks (which one student calls "more like clubs than they are cliques") takes place at weekend football games, church events, retreats to the beaches or mountains, and the bars and concerts of Winston-Salem.

"Students who sit in the corner and don't say much may have done fabulously in high school, but at Wake Forest, participation counts," says a senior. "Many professors count participation somewhere between ten and fifteen percent of a student's final grade, so it helps to speak up. The students I see succeeding are those who 'go beyond the call of duty'; they perform well, they visit their professors when they need assistance, they build relationships with classmates, advisers, and professors, and they make an effort to be

seen and heard. By taking that extra step, the opportunities I have had at Wake Forest have been endless." Another student says, "There are three common characteristics of students who do well. Those who work hard, frequently spending up to six or seven hours on homework a night, will do well academically at Wake Forest. Students who value honor and a tradition of integrity will do well personally at Wake Forest, and students who are willing to volunteer of themselves and their time will do well socially and emotionally at Wake."

Says one student, "In general, the people who are happiest tend to be well-rounded. The 'typical' Wake student could be described as motivated academically, involved extracurricularly, interested in others, works out fairly often, has a hidden talent or two (guitar, skiing, flying, etc.), and enjoys spending time with friends. Students also tend to be friendly—it's not unusual to be greeted by the smiling 'hello' of a stranger as you walk across the quad. Socioeconomic status varies greatly, but I am continually impressed and grateful that, in general, money and status do not influence attitudes or relationships. Though many students place a priority on their studies, they are generally willing to help another classmate better understand course material, and would gladly take a study break to catch up with a friend in Shorty's."

The honor code at Wake is a real part of daily college life, where, one student notes, "From freshman orientation to graduation day, students are reminded of the strong and heavily enforced code. They sign tests to ensure honesty and individual work, and questionable events are placed before a strict judiciary system. This strong code has helped

Wake become the reputable school it is today." Wake supports its students in many ways, including the Writing Center, learning disabilities support, and support for multicultural students.

A senior raves about the opportunities available for students at Wake: "I had the privilege to attend parts of the Lilly Conference on Religion and Higher Education in 1997, and during the lunch session, I was seated with the provost of Wake Forest and two of the most important voices in the discussion of religion and higher education. The conversation was amazing, and here I was, a sophomore, next to individuals who can change the way universities across the South view religious education. I am amazed that the opportunities continue. Since school has recessed for the summer, I received an invitation from Senator Pete Domenici and the vice president of the United States to attend a conference on ethics in government. I have attended numerous conferences like this one that have enabled me to network not only with students from around the country but also with leaders from around the world, all because Wake Forest believes in investing in their students." One student notes the impact of one of Wake Forest's many student-faculty and internship programs on her educational experience. "The STARS [Student Technology Advisors] Program has been incredibly beneficial to me because it is a paid job, it has given me technological training, I have had the chance to work with faculty and administrators, and I have acquired internship opportunities, all of which have better prepared me for the workforce. STARS pairs technically experienced students with professors to teach

all levels of computers so that the professors can integrate more technology into their classrooms. I have had the chance to work with several professors, as well as the vice president of finance and administration; I was placed in an awesome summer internship last summer; and I received training, paid for by the program, in Networking Essentials, Supporting Windows 95, Access, and Windows NT, with more training to come. STARS are paid by the hour and they work together to learn more about the IT field, all in the name of improving classes. The program has contributed to Wake's 'wired' reputation." As a senior eloquently puts it, "Charles Kuralt, the renowned recorder of Americana, reminds us that unique does not mean rare or uncommon, but in his words, 'alone in the universe.' I think that Wake Forest is unique in higher education. With its strong emphasis on technology, community, and opportunity, Wake Forest provides students with the tools and the resources to achieve great things, for Wake Forest educates students; it does not indoctrinate them."

What Happens After College

Students enter Wake Forest with postcollege goals very much in mind. One-third of every graduating class immediately enters professional or academic graduate schools. Many more will earn an advanced degree over time. Not only business school graduates but also a majority of arts and science majors move directly into the business world. Career choices include accounting, financial services, consulting, education, technology, management, sales and marketing, social services, and others. The college brags that "Wake Forest accounting students had the second-highest passage rate in the nation on the CPA exam and accounting majors are heavily recruited. Last year, 64 percent of Wake Forest students who applied to medical school were admitted, compared to a national acceptance rate of 42.4 percent." Some 42 percent of graduates head out of the Southeast for jobs or graduate programs, while a quarter are placed within the Southeast but outside North Carolina. Graduates have received a good share of the most prestigious fellowships, including seven Rhodes Scholars in recent years. The university enjoys high rates of alumni giving, upward of 40 percent.

WASHINGTON AND LEE UNIVERSITY

Lexington, VA 24450-0303
(540)463-8710

www.wlu.edu

Number of undergraduates: 1,680
Total number of students: 2,000
Ratio of male to female: 57/43
Tuition and fees for 1999–2000: $17,105
% of students receiving financial aid: 29
% of students graduating within five years: 87
Endowment: $769,000,000
Endowment per student: $384,500

Overall Features

What most defines Washington and Lee is its sense of history and enduring traditions. The college was founded in 1749 as an independent, nonsectarian liberal arts college to educate young men of social standing from the South for positions of leadership. Like so many of the eighteenth-century independent colleges established more on the ideals of the enlightenment than on sound financial footing, Washington and Lee was faced with financial disaster early on. The college authorities appealed to George Washington, who made a generous gift in 1786 that saved their tenuous institution from extinction. A century later the college suffered the ravages of the Civil War and badly needed a significant figure to lead its rebuilding. What more inspiring figure could be found than General Robert E. Lee who, as president, led the college to a position of strength and preeminence until his death in 1870.

Washington and Lee is committed to remaining a small college in order to cultivate strong personal relationships between students and faculty and among the students. There is a distinct effort to foster a culture of friendliness and respect for one another. Several of the distinctive features that influence the life of the school are the honor code, which governs all campus academic and social behavior, and "the speaking tradition," an invention of General Lee that requires that students, faculty, and administrators greet and speak to one another whenever they cross paths on the campus. Students express considerable respect for both the speaking tradition and the honor code as the guiding rules of expected behavior for the entire community. Students take exams unsupervised and their word is respected in all matters. The honor system is run completely by an elected student committee. All cases of violation are heard by the committee, which makes determinations on the rightness of the case and the punishment to be handed out. This is a major responsibility since the college operates on a single violation rule. The student panel is therefore required to vote for expulsion if a student is found guilty. It is important to an understanding of the college to appreciate the role these two customs play throughout the community.

The trustees decided to open the gates to women only in 1985 after much deliberation. Conservators of the institution's traditions and sense of community, they wanted to be certain that this would not lead to radical changes. The consensus of students, faculty, and alumni is that Washington and Lee has maintained its special ambiance while improving the academic qualifications of the student body and the quality of life. The women who enroll seem to be as positively influenced by the cultural codes that affect the school as the men are.

Washington and Lee's rural location was an intentional decision by its founders, who wanted to create an intimate community of students and faculty and protect students from the distractions of urban life. Little has changed in this regard over two and a half centuries. Faculty are hired both to teach and support their students through small class sections and interaction on the campus and in the dormitories. Faculty are evaluated on the basis of their teaching more than on their research and publishing. The location of Lexington definitely engenders communal spirit and strong personal relationships. Here is a particular story that illustrates the bonding among students and academic leaders that distinguishes the Washington and Lee experience: A group of graduates about to celebrate their fiftieth reunion from the college gathered to determine an appropriate class gift to the college to recognize their gratitude for the education they had received. They decided to establish a scholarship to honor the man who had been dean of students during their four years on campus, because of his commitment to the students and the college. The dean had been a mentor and friend to many of those who could remember, fifty years later, his impact on their young lives. This class determined they would raise $4.5 million to endow a scholarship in their dean's name for students whose character, citizenship, and activities made them desirable as campus citizens and who could not otherwise afford to attend Washington and Lee. This is a marvelous portrayal of the essence of the college's values.

Washington and Lee is a highly selective institution, with rigorous academic standards imposed by the faculty. A strong emphasis is placed on developing superior writing and critical thinking skills throughout the curriculum. Unusual for a small liberal arts college, Washington and Lee supports both an undergraduate school of commerce offering concentrations in economics, accounting, management, public policy, and political science, as well as a school of journalism and a law school, along with the college of arts and sciences. The student body represents thirty-six states, eighteen foreign countries, and 10 percent to 12 percent of students are of color. The college is committed to broadening its geographic, cultural, and socioeconomic student body to achieve national and social diversity. Financial aid is generous and the needs of almost all candidates who demonstrate need is met. Thirty merit scholarships are offered annually to students of exceptional academic ability. However, the percentage of students on scholarship does not yet match that of the more socially diverse peer institutions in the Northern and Western regions of the country. The Southern ambiance, with all its appealing traits, for many still characterizes the college. There is not a highly charged engagement with the political, gender, or social issues that one finds on many of the more liberal campuses. Students are more focused on their social lives, studies, and athletics.

A student with an appreciation for the beauty of a collegiate campus and surrounding environs could not do better than Washington and Lee. It is a pristine, beautifully designed and maintained campus. The Greek system is central to the social life of the school, 80 percent of men and 70 percent of women being members of a fraternity or sorority. There is little interest, it seems, in changing the role they play in the community, which is not the case at many of the other residential colleges of comparable academic quality. In fact, the college points proudly to its "fraternity renaissance of 1988–1999" with its "$15-million commitment to support and strengthen the W&L fraternity and sorority system." Varsity athletes compete at the Division III level in twenty sports and the great majority of undergraduates participate in intramural and club sports. Washington and Lee has a long-standing reputation in intercollegiate debate and sponsors a number of newspapers and lit-

erary magazines. For a small college there is much to do on campus.

What the College Stands For

Few institutions are clearer or more direct in declaring their educational mission and philosophy: "Washington and Lee emphasizes personal honor and integrity. No one attends the university without becoming aware of new dimensions of honor and integrity. Accordingly, students are given a large measure of freedom in governing their own affairs and are represented by active membership on faculty committees. . . . The same code of honor that governs academic life guides personal life. Washington and Lee strives for intellectual distinction. Its steady purpose is to be one of the nation's great 'teaching' colleges. Research is encouraged as part of the learning and teaching process, not as a substitute for it, or as a way of determining promotion or tenure of its faculty. Ideally, the university believes teaching and research cannot, and should not, be separated. Washington and Lee fosters an academic environment in which both teachers and students constantly learn—in classrooms and laboratories, in private research, in the give and take of extemporaneous discussions. All students are taught by professors, not by teaching assistants or graduate students." This is as forceful a pronouncement of institutional commitment to a first-rate faculty and outstanding teaching with ongoing research as a means to better instruction as any institution has promulgated. The university also defends its historic independence to influence young men and women in its own fashion. "Free of any control of church or state, the university is dedicated to the democratic form of social organization, to the dignity of the individual, and to the ancient freedoms, particularly to liberty of the mind with its attendant right of inquiry. Hence, the university is free to chart its own course, consistent with the highest educational standards, its traditions, and its aim of service to mankind." There is no confusion regarding the objectives this small institution has set for itself and is determined to realize for a selective group of young men and women.

College officials are proud of Washington and Lee's history and reputation. Curricular development, strengthening of instructional resources, improvement of facilities, continued increase in financial aid based on gifts and endowment (rather than tuition), and increasing diversity are institutional priorities. Washington and Lee's uniqueness, according to one administrator, is found in "our honor society, our curricular breadth for a medium-size undergraduate institution, a law school as our only graduate program, top-rank in all programs, and two hundred and fifty years of educational leadership and service."

Curriculum, Academic Life, and Unique Programs of Study

The educational theme of Washington and Lee is to foster students' exposure to the broadest range of knowledge and human understanding beyond the limits of any particular special interest. "The aim of the curriculum is to free the mind, lead to understanding, create humility and tolerance, and afford a basis for continuing study and learning." The emphasis on a true liberal arts education is similar to that of all of the Hidden Ivies. The curriculum

stresses studies in the humanities, and the social and natural sciences, especially in the first two years of study, which will give students the exposure to then choose a concentrated field of study and a preparation for graduate school. To meet the goals and purposes of the university, students are required to fulfill a number of distribution areas. However, there is a large number of available courses in all areas to meet their interests while also fulfilling requirements. Although there is less freedom to create a highly personalized program of study than at some other Hidden Ivies, there are formalized programs of interdisciplinary concentrations and independent study and research opportunities that professors help a student to design and manage. The faculty is determined to help students develop the vital skills of independent thinking, clear analysis and problem-solving, and the capability to communicate their ideas in their writing. The schools of journalism and commerce make it possible for undergraduates to combine the advantages of a liberal arts education with more defined studies in one of these two specialized fields. The university sponsors a good number of off-campus internships and study-abroad programs as a function of its mission to provide a breadth of intellectual and cultural experiences for students.

Major Admissions Criteria

The admissions office stresses that the single most important factor in evaluating candidates is the quality of their high school course load and performance in these courses. Indications of stretching oneself intellectually by undertaking the more advanced-level courses, and taking key subjects like English, mathematics, history, sciences, and language for three or four years are significant factors in evaluating worthy candidates. The committee prefers students who have taken the highest level of a subject its high school offers, be it honors or advanced placement, to those who breeze through with good grades in less challenging sections. As one senior administrator identifies the key admissions criteria, they are "academic achievement and ambition, demonstrated leadership, characteristics of honor and integrity, and energy." Another points to "academic achievement, intellectual ability, and honesty." How does this translate into the makeup of an admitted class at Washington and Lee? Of enrolled students 88 percent will have taken Advanced Placement courses and a foreign language, 100 percent will have had a minimum of three years of lab science, and 91 percent will have had four years of math. Leadership as demonstrated through student organizations, social service organizations, and athletics are other significant factors in determining who is offered a place in the incoming class. Test scores are also taken into account with the mean of those admitted at the high end of the scale, but not as narrow as the range among the Ivy institutions. Because enrollment is so small, entrance into Washington and Lee is very selective, but also very personalized. All of the above factors will be used to determine who most merits admission.

The Ideal Student

Any prospective candidate to Washington and Lee should review its mission statement and the essential features governing the life on campus. The school's location, size, and most popular social and extracurricular options should determine if this is the right match. The conservative nature of the student body and the approach to learning and living in this small environment will suit the traditionalist who wants a very challenging educational experience, a close-knit community, a spirited athletic program, and the social life that the fraternities and sororities provide. Washington and Lee cares as much about the values it imbues in students as it does for the quality of the academic education. One must buy into this system to grow and prosper over the four-year experience.

The school offers "a *total* commitment by faculty and administrators to support students," and maintains "a balance between high academic standards/superb teaching, and a spirited social environment." While not the most diverse of the Hidden Ivies, Washington and Lee has begun an effort to open up the university to more types of students. "In the fall," says an official, "a special trustee committee will explore ways to make W&L a more inclusive community for groups— racial/ethnic minorities and international students—that have not been well-represented in our traditional student population. This initiative is the *number one* priority of the dean of students' staff."

Student Perspectives on Their Experience

"I chose Washington and Lee because of the beautiful campus, excellent academics, social atmosphere, and sense of community," says one, characterizing the general sentiments of his fellow students. A classmate adds "the honor code and academic reputation" to the list. The university does a good job of "making each student feel like much more than just a number," and "puts on a helluva show when it needs to. W&L knows how to throw a party, official or unofficial."

The university could definitely attract a more diverse student population and expand the social scene beyond the Greek system, say some students. But others praise "the fraternity life. I've made the best friends of my life." And the honor code "is like no other anywhere in the world." "While it's not perfect," says a senior, "one of the most important things to me in my college career has been sitting alone in a deserted academic building at four A.M., taking a test with my books across the room, *on my honor*."

Those who do well are "extremely intelligent as well as socially compatible." And the faculty members are clearly the primary support structure for students. Says a senior, "The fact that all professors are so easily accessible helps students to succeed. You become very close with your professors and have the feeling that they can help you with many things." Another adds, "The professors many times go the extra mile to help out. They know their students. Ultimately, however, it comes down to God-given genius or work ethic."

Students who come to Washington and Lee should be prepared for a social scene dominated by fraternities, sororities, and athletics, a small college where "everybody knows everything," and a beautiful, challenging, and supportive campus and academic environ-ment. "During weeks in which you sleep ten hours in order to prepare for the barrage of tests, quizzes, projects, and papers," says a senior, "sometimes you find yourself alone on the majestic front lawn at sunrise, and you find peace with what you're doing."

What Happens After College

Reflective of the trend today of most selective college graduates, about one-quarter of Washington and Lee's seniors go directly to graduate school. About a third of these enter academic graduate programs, a third law schools, and a fifth medical schools. Of the three-quarters of a recent graduating class that went into employment, over one-fifth went into business and one-fifth into banking/finance. Other prominent fields of interest were consulting, communications, accounting, and education, with less than 5 percent of students each in the categories of retail, science, government, associations/non-profits, health and welfare, paralegal, and marketing. Well over half of alumni will enroll over time in graduate programs in mainstream fields.

WASHINGTON UNIVERSITY

Campus Box 1089
One Brookings Drive
St. Louis, MO 63130-4899

(314) 935-6000

www.wustl.edu

Number of undergraduates: 6,328
Total number of students: 12,035
Ratio of male to female: 50/50
Tuition and fees for 1999–2000: $23,634
% of students receiving financial aid: 60
% of students graduating within five years: 85
Endowment: $3,500,000,000
Endowment per student: $290,818

Overall Features

Washington University was founded in 1853 as an independent, non-sectarian institution to train men and women in the liberal arts and professions following the model of other major universities of middle size. The university is a combination of an undergraduate liberal arts college and four specialized schools and a graduate complex that offers degree programs in the professional fields of business, law, medicine, art, architecture, engineering, social work, and a host of academic disciplines at the doctoral level. Unlike the traditional small, strictly undergraduate colleges of note, Washington provides an extraordinary range of academic studies for undergraduates because of its size and the presence of the special schools. Within the undergraduate population of 5,400 full-time students, over one-half study in the college of liberal arts. Another thousand are enrolled in the engineering school, and the rest concentrate in architecture, art, or business.

The university has an astonishingly large endowment that places it in the top ten universities in the nation. Only three of the Ivy League institutions enjoy larger financial resources to carry on their educational mission. As a consequence of its wealth, Washington can function as a mid-size university with first-rate professional schools and a traditional college of arts and sciences that is the heart of the educational enterprise. Students have access to rich academic offerings, multidisciplinary concentrations, and independent study and research opportunities under the aegis of teacher/scholars. Several thousand individual courses are available to liberal arts majors that makes it easy to meet the distribution requirements. More specialized majors must undertake a number of courses in the arts and sciences. Washington is a research university, and faculty members are expected to carry on their own scholarship, but they are also expected to take their teaching responsibilities seriously. The overall environment is moderately traditional since the great majority of students are focused on their studies and future career plans. There is good variety in the backgrounds of the student body, who come from all fifty states and eighty foreign countries. Of undergraduates, 86 percent are from out of state and 25 percent are of color or international background. Thanks to $40 million allocated to undergraduate scholarship funds, the university provides financial aid packages to 60 percent of its students. The average first-year award is $19,000 and ranges up to $28,000 per year. Washington is able to fulfill its goal of attracting deserving students of all cultural, social, and economic backgrounds who qualify for admission. A large number of merit-based scholarships are also available to attract outstanding students to the various undergraduate schools.

The campus, located in a suburban section of St. Louis abutting a 1,300-acre park and the famed St. Louis Zoo, is handsome and well-equipped. The majority of students live on campus in one of twenty-two residencies, while a small percentage live in fraternities on fraternity row. Only one-quarter of men and women join a fraternity or sorority, so their influence on the social life of the campus does not overwhelm the many other social organizations and student activities supported by the university. There is a good sense of a campus community considering the size of the school and the urban location. Students are fortunate to have a wealth of activities available to them both on campus—from special interest clubs to myriad performing arts groups—and in the cosmopolitan city of St. Louis. Music, theater, art museums, restaurants, and professional sports teams are easily accessible to students. Although Washington's leaders determined years ago to shun the world of intense intercollegiate athletic competition that dominates the Midwest university world, there is a strong Division III program of competition in seventeen sports for men and women. Washington helped to found the University Athletic Association of nine institutions similar in size, academic quality, and student body. They include such sister schools as Johns Hopkins, Carnegie-Mellon, Brandeis, Rochester, Case-Western Reserve, and the University of Chicago. Thanks to excellent athletic facilities, 75 percent of undergraduates participate in intramural sports. Many students who are attracted to Washington for its diverse academic programs, student body, and urban location like the opportunities to get involved in community service. This is clearly reflected in the large number of social and service organizations that students can join. The diversity of the students is also reflected in the many affinity groups that represent virtually all of the gender, racial, ethnic, and political concerns alive on most campuses. Yet the atmosphere is not charged with intense activist attitudes or behavior. One undergraduate commented that his peers have personal convictions and passions on various issues, but do not press them on others in the in-your-face style popular on some other campuses. It is rare to hear a student lament that there are not enough things to do or academic offerings to choose from. The more pertinent issue is how to choose from such a cornucopia of options in all respects while staying on top of the demanding workload. Washington students are concerned about their grades since most are headed to graduate studies but they also value downtime and engagement in clubs, sports, or the arts.

What the College Stands For

Washington declares that its primary mission is to promote learning, both of its students through classroom studies and its faculty through research and teaching. As a liberal arts college and research university, Washington has a dual objective of educating the beginning learner through training in the arts and sciences, and at the same time, adding to the store of human knowledge through extensive research programs in the humanities, sciences, medicine, and technology. For undergraduates, the goal is to "prepare students with the attitudes, skills, and habits of lifelong learning and with leadership skills, enabling them to be useful members of a global society." The model graduates will be bold, independent, and creative thinkers who will give back to the larger community.

Washington's identity is wrapped up in its Midwestern location and ethos, undergraduate and graduate balance, and residential environment. Says an administrator, "During the late 1950s and the early 1960s, Washington University decided that it was going to become a national—indeed international—university. Our leaders envisioned a world-class university located in the Midwest as an advantage for our

students and for the region in which we are located—a place where collaboration, interdisciplinary learning, excellence, hard work, and success are basic attributes. Add to that our Midwestern qualities of friendliness, inquisitiveness, and respect for individual differences. The educational process here operates within the framework of these qualities and attributes. For our undergraduates we provide an experience that touches on virtually every aspect of a young adult's transformation from high school and home to becoming a mature individual who has developed a strong grounding in world cultures, literature, the sciences, the humanities, and our socioeconomic frameworks. That transition occurs well beyond the classroom, laboratory, and studio. It happens as part of a student's residential life experience, philosophical and spiritual growth, and a sense of purpose that distinguishes a well-rounded learner."

A detailed set of educational goals for the university's students, prepared in 1996 by William Glass, a professor in the humanities, begins by stating, "Education should be lifelong and lifewide. You may leave college, but you ought not to leave learning." Glass goes on to set out Washington University's specific goals in two major areas, illustrating the breadth of the liberal arts education that the university seeks to provide: personal development (set high standards, know thyself, solidify good character, be ready to accept social and environmental responsibility, and recognize that the arc of learning is lifelong); and intellectual development (reasoning, communication, aesthetic appreciation, mathematical skills, historical perspective, scientific understanding, literature, international and cultural awareness, and information acquisition and research)."

Washington University's endowment has not only allowed it to pursue, enroll, and retain talented and needy students and top faculty, but also to continue to develop its Frederick Law Olmstead–inspired campus infrastructure. "We are still creating our campus," says an official. "Currently, more than $225 million in construction is either underway or planned in the short term for the Hilltop Campus. This includes many new residence halls, refurbishing and rehabilitating numerous original buildings, and construction in recent times of a new athletic complex, a new home for the school of business, several new facilities for the sciences, a psychology building, executive education center, and many more. Currently on the drawing boards are a new visual arts and design center for the schools of art and architecture, as well as for the gallery of art, art history department, and the art and architecture library. Also soon to start are a new university student center, yet another science building, and many more improvements in existing buildings." Did we mention that the university has a lot of resources?

Curriculum, Academic Life, and Unique Programs of Study

The presence of a large teaching faculty, an extensive curriculum in the arts and sciences, and the professional schools of visual arts, architecture, engineering, and business lets students develop almost any course of study of their liking. The university encourages students to combine studies in different fields within the arts and sciences and between the liberal arts and the professional programs. A few examples are programs in the biological sciences and medicine, architecture and engineering, and performing arts and business. Washington is an excellent place to explore potential long-term intellectual and career interests before making a final commitment. As an administrator says, "We want to remain a medium-size campus with the resources and services of a much larger university, but with the supportive, comfortable atmosphere of a small college. We also want our students to prepare for the twenty-first century by designing programs that draw from more than one discipline and also provide excellent, in-depth preparation for careers and for life."

Almost two-thirds of the most recent graduating class at Washington earned both a major and minor, or more than one major. More than one-third completed requirements for two majors and some students earned two degrees from two different schools at the university. "These choices," says an administrator, "allow students an intensive focus on one or more subjects, uniquely combining such areas as engineering and art, business and Russian, architecture and history, computer science and visual communications, history and French, or English literature and management, etc. It is even possible for students to earn a bachelor's degree and master's degree in one or more disciplines under several combined programs." Undergraduates do have specific distribution areas to fulfill but have great flexibility in doing so. The university encourages undergraduates to take advantage of its sponsored study-abroad programs that involve study in more than just languages and the arts. The Olin Library and departmental libraries provide undergraduates with outstanding facilities for study and research. The faculty incorporates a good deal of research and self-directed study for students within a formal course curriculum, and students are encouraged to partner with their professors on research projects. The theater, art museum and studios, and science facilities are considered outstanding. There are sixteen specialized centers for research and study on campus that can appeal to a student who has a highly focused interest in one of the fields.

Major Admissions Criteria

Admission has become more selective in each of the past several years. In 1998, 16,000 high school seniors applied from all regions of the country for a first-year class of 1,475 men and women. The growing popularity of the university due to wider recognition of its academic excellence, considerable resources, and quality of life has attracted increasingly high-performing students. Says the college, "The word is out among high school students that Washington University is a great place to go to college. This is the result of genuine, word-of-mouth communications traveling from satisfied graduates to high school students and their parents around the country. Applications have quadrupled over the past dozen or so years, and they have more than doubled in the past four years." Washington University has an excellent admissions and enrollment process, involving personal letters to prospective students, merit awards to talented admitted students, and early identification of potential applicants, which has also contributed to the university's higher vis-ibility outside the Midwest. Almost all of those who enrolled in the most recent class graduated in the top quintile of their high school class, took the most enriched courses available to them, and presented high test scores. The generous financial aid policy helps greatly in attracting many outstanding candidates of diverse backgrounds.

What is the admissions committee looking for as it sorts out such a large pool of qualified applicants? Beyond the statistical profile of grade point average, test scores, and class rank, it is looking for evidence of a serious commitment to learning and a particular talent or activity that indicates a student will make a special contribution to the community and take advantage of the many resources the university has to offer. Officials say: "Washington University seeks students with a range of qualities that we believe will contribute to their success in life:

• The student should pursue a challenging curriculum in high school where it shows he or she has received strong academic preparation as well as demonstrating strong performance.

• It is important that students have special talents and character—factors that can be revealed through the application for admission, the required essay, recommendations from teachers and others, or through the optional interviews offered with our alumni in many areas of the nation and the world.

• Successful applicants also show a commitment or passion over time for one or more extracurricular activities and/or a job or other responsibilities.

Strengths in each of these three areas are important for the successful applicant."

The Ideal Student

Men and women who are serious about their studies and who want to grow in other dimensions as well will enjoy their experience at Washington University. The range of opportunities both intellectual and social and the campus-centered urban setting works well for the adventurous individual who is looking to stretch his or her experiences beyond the home and school environment. Those with a natural curiosity to explore new activities and make new and differing friendships while balancing their academic obligations will gain immensely from their four years of living at Washington University. While it is not a huge institution, it does take individual initiative to profit from what the community has to offer. One cannot sit back and expect teachers and advisers or other students to automatically reach out. It helps to have some idea of interest areas in order to select from the comprehensive range of academic opportunities.

As an administrator notes, "We seek students who can thrive here, who are self-starters, and who look forward to sharing their college experience with people from abroad and different backgrounds. We want students who can explore their own potential to learn more about themselves and their talents. Our professors make themselves accessible both inside and outside the classroom—in fact, virtually all courses are taught by regular faculty. These faculty are dedicated to the success of their students and at the same time lead in their research fields, write definitive books,

create works of art, and win major awards. They embrace teaching as an integral part of their commitment to the university and to its students." Classes at Washington are larger than at most of the smaller colleges in the Hidden Ivies, but are still small relative to most universities. Students who thrive at Washington enjoy these more intimate class settings and the opportunities for research work available with faculty, but again, have more personal motivation to seek out support and contacts. "More than three-fourths of Washington University's undergraduate classes range between one and twenty-four students," say college officials. "Where classes sometimes may be larger at first, they generally become smaller as the student progresses into his or her chosen field or fields of study. And for those students who seek direct interaction, many have found that they can work shoulder-to-shoulder with faculty on research and scholarly projects outside the traditional classroom." Students are assigned advisors to help them along their path, and they find a wide range of athletic, social, and residential life experiences to choose from. What students should not expect at Washington is too much hand-holding or an environment where everyone is studying the same subjects or is interested in the same social outlets.

Student Perspectives on Their Experience

Students choose Washington University for its diverse academic offerings, academic reputation, friendly urban/suburban community, and sense of balance. Says a senior, "Wash U had a great academic reputation and offered so many different majors and possibilities. I made my final decision when I came to visit the school—the community here is so friendly and helpful. You feel like yourself here, not a number. The deans, professors, and faculty are all very accessible and they want to help you succeed. It felt comfortable when I visited and I knew this was where I wanted to be for the next four years of my life." A sophomore says, "I chose my college because of the people who work, teach, and go to school here. Sure, it's easy to read magazines and reports that give statistics and rankings of schools, but once you get to college none of that stuff really matters. It's going to be the teachers, the students, and the people that you come into contact with that impact you the most. You have to be able to see yourself as a part of them in order to feel comfortable and receive the best education you can."

An architecture student adds: "Washington University has a diverse selection of attributes and educational opportunities to complement its successful architecture program. A large part of Washington University's immediate appeal is the fact that the institution sends out a lot of mail and is very interested in all applicants. I chose WU because of its larger urban area location and its beautiful campus (in old Gothic style with pink granite). WU also had the educational opportunities and a higher caliber of students and faculty that I was looking for." A sophomore says, "I chose Washington University because it provided 'big university' resources while maintaining 'small university' friendliness and faculty involvement. On top of having incredible resources, the campus is gorgeous and conveniently located."

The university "offers millions of opportunities: internships, study abroad, extracurricular involvement, sports, work, majors/minors, everything. The possibilities are endless. If you want to do something, the school will do its best to help you. The school also does well with keeping up buildings and staying on top of technology." Another student says, "What Wash U does best is take care of its students. It doesn't just put up a show for the brochures and magazines. There are so many resources for students, whether you need help with classes or are looking for a job or want to get research experience, the people at Wash U are there to help you. All you have to learn to do is ask. I was taken aback at just how willing to help everyone was. I wasn't used to that kind of treatment, but I realized that by taking full advantage of the opportunities presented to me I would become a better person."

Those who do best "keep up with their studies, have an interest in the subject matter, develop relationships with their professors, and seek help when needed," says a senior. A sophomore echoes this: "Everyone is smart at this school. It's how hard you work, how you use resources, and how much you concentrate on priorities. Slackers do not get very far at Wash U. Many students who breezed through high school by cramming for tests the night before have to change their study habits to survive here."

Students feel supported by faculty, friends, and the extra resources available on campus. "The staff at Student Educational Services is wonderful, they provide tutors, organize study groups, help with time management, some of everything. They don't just get tutors who are knowledgeable about the subject matter, they hire tutors who are knowledgeable and who want to help others understand. The tutors are excellent in explaining difficult concepts in easier terms. Also, the support from academic advisors and peer advisors helps us to succeed, too." Another student adds, "Students have a lot of help in succeeding here. Though it's a very challenging school, the attitude is not one of competition but of camaraderie and group learning. Those who need a little extra help can still succeed when they take advantage of TAs, office hours of the professors and tutors, as well as their fellow students. Our motto is 'Play hard after you work hard.' Everyone here likes to have fun, but everyone is also very academically oriented. It helps a person to work hard when everyone else around them is doing the same."

The university is seen as containing a diverse mix of people. Says a senior, "Our campus is very diversified in accordance with culture, ethnicity, race, socioeconomic status, and other factors. There are various cultural organizations that are very much involved in making the campus community a comfortable one for all students. The university also provides diverse programming, whether music, lectures, or theater performances that touch on many different groups of people." Another student reflects, "WU is a multicultural melting pot. It is probably one of the only schools in the nation where students of all races, ethnicity, backgrounds, etc. get along like friends. By living in the residential college, students get to experience different cultures every day. For example, my floor contained fifty students—twenty-five male, twenty-five female; three of these were African Americans, eleven Asians, three Hispanics, six Indians—all from five different countries

including the U.S. Now, I just had to think hard about this, because we all became such good friends. By living together, we learned to put behind us the colors of our skin and we were able to see the true individuals that we are. That experience alone was almost as valuable as the education."

The only critiques of Washington seem to center on the need to improve communication between the various parts of this university, among the different schools and programs, and connecting academic and residential life. Students also mention the large amount of construction on campus (Remember all those new buildings the university has planned?), and the continuing development of the residential-life system. This is something the university sees as positive, reflecting the increasing number of students wanting to live on campus after their first year and through their tenure at Wash U.

"Washington U is a great place to be," a senior states. "It offers so many opportunities and the wealth of knowledge contained here is phenomenal. Don't let the price tag of attending turn you away: Need-based and merit-based financial aid is available and financial services will work with you to help find a price your family can afford."

What Happens After College

As already noted, a majority of graduates continue their academic career in graduate school either immediately upon graduation or after gaining work experience for several years. Some 84 percent pursue advanced studies within seven years of graduation. There is great diversity in the fields of study graduates pursue—they include not only the traditional professional schools but also the arts, architecture, sciences, technology, education, and social work, among other disciplines. If a student performs well in his or her undergraduate studies, he or she will have many opportunities for graduate work thanks to the university's reputation for rigorous academic preparation. A good number of recent graduates have received competitive awards that include the Rhodes, Fulbright, Marshall, Truman, Mellon, National Science Foundation, and Beinecke Fellowships. Acceptance rates into the professional schools, particularly medicine, science, business, and engineering, are very impressive. Alums are increasingly loyal and, according to the university, report an "extraordinarily high satisfaction rate with their experience here . . . In the last two decades, participation rates in the annual fund have more than doubled."

WELLESLEY COLLEGE

106 Central Street
Wellesley, MA 02181-8292

(781) 283-2270

www.wellesley.edu

Number of undergraduates: 2,175
Total number of students: 2,175
Ratio of male to female: 100% female
Tuition and fees for 1999–2000: $23,320
% of students receiving financial aid: 52
% of students graduating within five years: 85
Endowment: $780,872,000
Endowment per student: $359,021

Overall Features

Wellesley's administrators, faculty, and trustees are quite clear on the priorities that define the nature and purpose of the institution. The college was founded in 1876 by Henry Fowle Durant, a prominent educator who believed strongly in the right of women to have access to a liberal arts education comparable to that which men received. There is no questioning the quality of the education Wellesley women have obtained over the course of the college's long history. After considerable discussion during the 1960s and 1970s, when all of the great New England men's colleges began to admit women and several of Wellesley's sister institutions decided to admit men, the trustees decided to remain faithful to Wellesley's original purpose of providing young women of talent and intellect with a distinguished education in an environ-

ment that put women's interests and issues at the forefront. Wellesley has maintained its excellence through these latter years of change and foment in higher education. No other liberal arts college in the nation can claim to offer stronger intellectual training or a more complete and beautiful campus on which to live and learn.

Having determined to remain in the forefront of educating only women, Wellesley's leaders and faculty include in their teaching and campus activities a women's agenda of significant issues. They promote the fact that every dollar spent on teaching, facilities, research, residential life, student organizations, and athletics is for the purpose of advancing the schooling of the women under their tutelage. There is one teacher for every ten students and all of them are engaged in teaching and advising. The administration takes pride in the equal balance of men and women on the faculty. In the opinion of the dean of admissions, Wellesley has retained what has always been the best features of its education while developing major new directions and approaches in curriculum and teaching. Wellesley's leaders are convinced that the experience of associating with talented, curious, independent peers of many different backgrounds and a preeminent faculty is at the heart of the educational experience. Thus the residential college that is very selective in creating its community each year is considered a matchless experience for personal development and intellectual growth.

Wellesley is better able to carry out its goals than most liberal arts colleges today, thanks to its robust endowment, magnificent campus and facilities, and the passionate support of a majority of alumnae. Recent examples of their generosity include the new Technology Center, described as the pulse of the campus with its many advanced technological features for learning. All college buildings are wired to the Center, which makes possible easy communication among all members of the community, as well as advanced levels of research and teaching. Wellesley believes that its graduates must be literate in technology if they are to excel and lead in their chosen endeavors. The Teaching and Learning Center is also a new addition, providing academic support to students and training for faculty to learn how to teach their subject in the best possible manner. The college's wealth allows the admissions committee to admit any qualified applicant without concern for her financial need, which results in a highly diversified student body of high academic quality. Over 50 percent of students

receive financial aid today. The trustees, again in keeping with the goals of the college, recently added $2 million to the financial aid budget so that students' financial aid package would include more grant money and fewer loan obligations. As a result of this action, debt loans have been reduced from an average of $20,000 to $12,000. This puts Wellesley in the rarefied company of such preeminent institutions as Princeton, Harvard, Yale, Dartmouth, and Stanford.

What has changed most in the past several decades is the composition of the student body. Wellesley women were, for most of the college's history, advantaged white women from economically comfortable, traditional backgrounds. At present 40 percent of the women are from multicultural backgrounds. This is what gives the community its energy, diversity of opinions and cultures, and thus provides a growth experience for every woman in residence. The twenty-one residential halls are attractive, comfortable, and self-sustaining. Students are assigned randomly to a residential house for their four years to ensure exposure to a mix of students. There are dining rooms in the houses, apartments for faculty masters who are responsible for the advisory system, and cultural activities that have generous support from the college. Wellesley operates on a system of democratic government for the community; thus students have many opportunities to participate on committees and advisory boards that determine school policies and programs. Extracurricular activities take many shapes and forms. Theater, music, art, community service, political clubs, newspaper and literary magazines, and special interest groups make up the 140 organizations that are recognized by the college. Wellesley sponsors intercollegiate teams in all of the major sports in addition to dozens of intramural teams that compete among the residential houses. The new sports center is an impressive modern complex that includes all of the facilities necessary for an outstanding athletic and personal training program. The crew program has a long and special tradition at Wellesley, with excellent facilities for training.

What the College Stands For

Wellesley's leaders sum up their belief in the importance and value of a liberal arts education in this way: "In a time when technology is widely accepted as a substitute for ideas, when action all too often preempts thought, and when instant analysis masquerades as understanding, a liberal arts education has never been more crucial. In such a time, Wellesley College may well offer more to its students than ever before in its history." What lies behind this pronouncement is the belief that the ethical and moral values developed over the history of mankind are essential guideposts in making critical judgments in our complex society. Wellesley faculty and advisors stress this concept throughout the curriculum and the residential life in which all students participate. The college uses its financial resources to admit outstanding women of all backgrounds, because they believe that exposure to a highly diverse community is one of the most significant elements of the learning experience at Wellesley. Another mission of the college is to play a leading role in advancing the interests and needs of women in contemporary society as well as to instill a general sense of social

responsibility. These concerns permeate the academic and social life so that students will feel they can express their beliefs and will have the confidence to achieve their goals in the future. Self-governance also weaves its way into the fabric of the Wellesley philosophy. The more opportunity for young women to make daily decisions for themselves and govern their own behavior, the better prepared they will be to lead independent and fruitful lives.

Another element of college life that takes on particular importance is Wellesley's honor code: "Throughout its history Wellesley College has based its student life policies on the concepts of personal integrity, respect for individual rights, and self-government. Established in 1919, the honor code rests on the assumption that individual integrity is of fundamental value to each member of the community. The honor code requires all students to display respect, responsibility, and honesty in all aspects of college life. The honor code provides students with privileges that are unique to the college. They include self-scheduled exams, open-stack libraries, guests in the residence halls, and full use of college resources. The fact that these privileges have been instituted and continued without major incident for eighty years is a testament to the respect Wellesley women have for the honor code and for one another."

In the future, Wellesley aims to continue its academic strengths and traditions while ensuring its relevance for its students in a changing society. As an administrator notes, "Wellesley College will continue to adapt as needed to respond to the rapid globalization of the world's economic, political, and cultural systems. The people who will effectively manage these new global realities will be those who can move from culture to culture, who can collaborate and communicate with fluency across national, racial, religious, and socioeconomic lines, with knowledge, sophistication, sensitivity, and ease. The work environment of the future will demand both intellectual flexibility and task-specific disciplinary knowledge. There is no better preparation for this environment than a Wellesley liberal arts education."

Curriculum, Academic Life, and Unique Programs of Study

Wellesley is committed to the tradition of the liberal arts education, which means that students will gain both breadth in the major fields of the arts and sciences and depth in a chosen discipline. As an administrator puts it: "The mission of Wellesley College is to provide an excellent liberal arts education for women who will make a difference in the world. The fundamental foundation to Wellesley's successful attainment of this mission is the intellectually serious, rigorous, and personal instruction that students receive from our faculty. Wellesley's faculty members cherish the opportunity to work closely with their students; in fact, more than four hundred students each year conduct research with their professors. Our strong liberal arts curriculum is broad, with more than nine hundred course offerings, and deep, with offerings from introductory classes to advanced seminars. Wellesley's distribution requirements ensure that each student's liberal arts education is built on a strong foundation of core knowledge. The goal of these requirements is

to expose students to a wide range of issues and ideas that will broaden their knowledge and worldview while providing the flexibility to pursue personal interests and curiosities."

The faculty recently undertook an intensive internal review to determine what skills are necessary to become an educated person and a leader in the twenty-first century. The results of this study serve as road signs for the college to navigate the academic curriculum and to plan for the constantly fluid world that awaits its graduates. As one senior administrator expressed it, the end goals are to have students seek the major truths of life and develop the tools to continue their search for the rest of their lives; to become good writers, avid readers and researchers for informed decision-making, and critical thinkers; to know how to solve complex problems; and to be able to articulate their ideas with fluency.

There are no core courses that must be taken, but there are distribution requirements to be certain that students choose widely from the exceptional number of courses available within the thirty-two departments and twenty-four interdepartmental majors Wellesley supports. It is safe to say that the breadth of subject areas and interdisciplinary programs offered are second to those of no undergraduate liberal arts college in the country. Wellesley has always encouraged its students to survey the sciences, which are considered among the strongest departments. The faculty encourages students to design a double major or opt for an interdepartmental major to fulfill their special intellectual interests. In order to broaden still further the learning opportunities for students, Wellesley has a full cross-registration policy with MIT. Typical courses taken by Wellesley students at MIT are technical sciences, architectural design, computer science, and business finance. Wellesley is also a member of the Twelve College Exchange Program, the National Theater Institute, and the Maritime Studies Program, among other cooperative arrangements. Students are encouraged to gain experience in other cultures through the many study-abroad programs that are sponsored or cosponsored by Wellesley. The school calendar includes a winter session in January that takes off the pressure of the regular semester academic obligations and allows students to undertake a special interest course, to work or intern off campus, to create and complete a special project, fine-tune certain skills, or do volunteer work. A new academic project called the First Year Experience Cluster is designed to give students who elect to participate an in-depth exposure to critical thinking, reading, discussion, and writing by focusing on a specific topic.

Major Admissions Criteria

The admissions committee is quite clear on the most useful information that determines if a candidate is likely to succeed in this heady intellectual and social community: the high school transcript, which is a four-year history of a student's curriculum, credits, and performance. The nature and quality level of the courses is the best indicator of intellectual curiosity, motivation, and the skills to meet the demands of the Wellesley education. The admissions committee is concerned that students do not overextend themselves, either. They look for signs of challenging oneself in advanced courses of major interest, but it is not critical to take nothing but Advanced Placement courses in the junior and senior year. Teacher recommendations that describe a student in depth are of great value to the committee since they are witness to the degree of intellectual curiosity, engagement in the learning process, maturity in thinking, and enjoyment of studies. Test scores do matter but there is no specific cutoff for consideration. As an official says, "The Board of Admissions seeks students who have intellectual curiosity, the ability to think, to ask questions, and to grapple with the answers. Wellesley looks for a student who is excited about learning. This comes through in the Board's review of her academic record, her recommendations from teachers, and in what she writes about herself in her application. The College also looks for signs of leadership skills, motivation, and commitment."

While Wellesley attracts its share of top academic performers each year, it can be somewhat flexible in its decision process because all of its available places go to women. Consequently there is great interest in the candidate who has achieved in a particular talent arena, be it music, drama, art, dance, athletics, or leadership. The dean of admissions summed up their policy this way: "Every decision made by our committee is an individual decision and the total person, not just grades or test scores, is taken into consideration." Wellesley is somewhat unique among the major colleges as it includes both elected faculty and junior and senior peer readers in the review and recommendation process for admission. Because its leaders believe so strongly that the heart of the Wellesley experience is the opportunity to live among talented, diverse, curious peers and outstanding faculty, the admissions community gives considerable weight to the kind of person who will enhance the residential community.

And again, needy students should take note that, as the college makes clear, "Wellesley College is one of the few most selective colleges and universities that has a need-blind admission policy and meets 100 percent of a student's demonstrated financial need. Applicants are considered for admission only on the basis of their talents and personal qualities, not on their ability to pay. If a student has financial needs, the college works with the student to establish a financial aid package that meets 100 percent of the demonstrated financial need, based on a financial aid formula. These policies mean that Wellesley is able to attract the best students, regardless of their financial resources. Keeping a Wellesley education accessible and affordable is one of the college's top priorities. We are fortunate to be in such a healthy financial position to be able to maintain these admissions and aid policies."

The Ideal Student

Similar to all of the colleges and universities in this book, Wellesley has high expectations for its students not only in the classroom but also in the life of the community. While students have a great deal of autonomy in choosing the courses they find interesting and creating a major field of study which meets their intellectual goals, they will be confronted with a very demanding level of work. Sloppiness in thinking, writing, and presentation will not be excused by the faculty. A premium is put on individual interpretation and

the ability to defend one's position on any topic. Wellesley women are also expected to carve out one or more activities from the huge block of opportunities and to excel in their elected endeavors.

Wellesley is definitely a women's environment. A concern for women's interests and challenges is also part and parcel of a student's commitment as far as the college leaders would have it. But for all of the advantages of the college's location in the Boston/Cambridge area and the cross-registration opportunities with other major institutions, many Wellesley women lament the lack of a natural and readily available social life. One has to reach out to the larger world beyond the gates of this semi-secluded campus to develop relationships with members of the other sex. While most women choose to attend Wellesley primarily because of the excellence of its education and facilities, a distinct cohort of feminists want little to do with men in the traditional dating scene. They find at Wellesley the opportunity to engage in active debate on the issues of feminism and the barriers to equal access to careers and leadership roles. For all of this, the prospective student should be more concerned for the distinctive educational and community programs Wellesley fosters than for the more traditional model of the coeducational collegiate experience that strikes more of a balance between intellectual and social components. Wellesley is also an appropriate place for the intelligent and inspired young woman who will feel more comfortable in a less competitive social environment, one that provides a good deal of support from the faculty and administration.

Student Perspectives on Their Experience

It would be rare to hear a Wellesley student complain about the quality of the education she has received. From the outstanding faculty and its interest in and support of a student to the curricular opportunities to the physical facilities and campus, Wellesley is as good as it gets. Most students express their appreciation for the many advantages Wellesley provides. They find that concern for their welfare is demonstrated by the many layers of personal and academic counsel and support, the financial aid that is generously dispensed, and the multitude of choices for learning on and off campus. As a senior describes, "I attended a very large high school (five thousand students) and I grew up in a mostly male household. Wellesley seemed to offer me the best of two worlds. First, it was an excellent small liberal arts college. Second, it is affiliated with a top research institution (MIT). I applied to fifteen colleges (Ivies, state schools, city schools, and coed liberal arts colleges) and was accepted to all fifteen. However, there was something about Wellesley that clicked with my personality. I knew I had found an environment I could thrive and grow in. Wellesley is a student-focused institution. There is something for everyone. Wellesley provides its students with a wealth of resources to aid in their education process. These resources are not only financial but include the people (faculty, staff, students, and alums). We have great relationships with our alums, who are always ready to help."

The main reasons some women are unhappy center on the lack of a coeducational environment; they lament the absence of men in class discussions and hang-out time. Discontent seems to be less about formal dating than about having men around as an option for social and intellectual engagement. As one woman puts it, "Women at Wellesley, in general, do very well in life. Wellesley is able to maintain its reputation as a place that breeds great women, women who will make a difference in this world. But this achievement comes at a price. Women at Wellesley give up a normal college experience—complete with men, partying, and intellectual/social activity—for one that resembles the experience of a women's boarding high school. Most women accept their life at Wellesley for what it is, and instead focus their attention on their studies and the dream of life after college. The knowledge that Wellesley is not 'normal' is very stressful and difficult to accept for many students, especially those who spend time with friends at other schools." Some students feel there is too prevalent an edge to feminism or diversity for them on campus and that this puts pressure on individuals to conform to an agenda or to be part of a particular group; other students say there is too much emphasis on work at the college. One first-year student reflected on her desire to transfer to a coed institution where her friends were enjoying themselves. Her dilemma, she told us, is knowing what she would be giving up in terms of programs and faculty if she were to leave Wellesley. Yet other students want the college to be even more involved in social change: "I would like my school to be less isolationist and more radical," says one. "I would love for Wellesley to be a school for rebels and activists." Wellesley applicants are advised to measure the campus environment carefully, to spend time at the college and talk to current students to make sure this school is the right match for them, because, if it is, it offers an amazing and empowering undergraduate college experience.

What Happens After College

Any alumna will attest to the value of her Wellesley degree in the larger marketplace and to the doors it opens to prove her abilities and drive. The Wellesley alumnae network has long been a powerful resource for jobs in a wide range of professional careers. The college maintains a well-organized information and advisement center for its students to connect with potential employers and meet the requirements for graduate schools of all kinds. As with the other top colleges today, a majority of graduates will explore jobs and new parts of the world before continuing their formal education in advanced programs. Here, too, the reputation for academic rigor is such that Wellesley women are very successful in gaining admission to top graduate programs. A college official says, "Throughout our history, Wellesley has prepared women to be engaged citizens and leaders . . . Among the college's 35,000 alumnae are a pioneering environmentalist, a construction engineer, a NASA astronaut, the first female U.S. secretary of state, journalists, artists, community volunteers, and government officials around the globe, to name only a few. Our alumnae recognize the value of their Wellesley education and remain tremendously supportive of and loyal to the college."

According to the college, "Wellesley's rigorous and well-rounded liberal arts education prepares its students for achievement in a wide variety of fields and a changing work environment. Approximately 20 percent of recent Wellesley graduating classes have entered graduate schools following commencement. The fields of graduate study for the class of 1999 are: 21 percent in education; 15 percent in medicine and other health-related fields; 15 percent in humanities and law; 12 percent in the sciences, mathematics, and computer science; and 5 percent in fine or performing arts, architecture, and design. Wellesley's acceptance rate to medical schools (71 percent in 1998) is significantly higher than the national average. In the past four years, more than 80 percent of all Wellesley graduates applying to law school have been accepted. The majority of Wellesley graduates each year pursue employment upon graduation. More than 130 top businesses and 40 non-profit organizations recruit on campus each year. For the class of 1999 about 22 percent anticipated careers in business or industry; 14 percent in education; 8 percent in government and public policy; 7 percent each in communications, arts, and medicine or other health-related fields; 5 percent each in law and social services; and 4 percent in computers."

FOUR
Words of Wisdom for Parents and Students
(FROM PARENTS TO PARENTS)

As college admissions time approaches, it is the rare parent who does not feel a wave of potentially debilitating emotions, from confusion, bewilderment, and impotence to varying degrees of panic and desperation. If these emotions strike a chord of recognition, you are ready to begin the admissions process! Let us encourage you to see the admissions search as a joyous occasion; that's right, a time for celebrating the opportunity that awaits you.

As you consider the many milestones in your child's life in which you have played a significant role already, you can feel confident that you will also meet this phase of his or her development successfully. Remember that you helped him to accomplish each of the previous milestones in good order, otherwise he would not be at this point in his journey to adulthood and independence in preparing to enter college.

Over the course of their high school junior and senior years, you will have the opportunity to communicate . . . okay, you will have to work hard to extract tentative ideas and concerns about the next phase of education from your children. They are counting on you to be available and supportive throughout the search, although at times they will resist any offer of help. This will be as critical a parental role as the early years of teaching them to take care of themselves. Imagine the opportunity you now have to support them in determining what they want their future to be, what skills and talents they can build on, and what kind of living/learning experiences they can grow from next.

Most parents who have experienced the college-search process can recount the good times they had visiting campuses around the country with their children. They remember how the kids would talk about everything while traveling in a car, even to them. Haven't you had the sense that the car was in some way a sanctuary or neutral ground where all topics small and large could be articulated? Wasn't this the setting in which you first learned of their triumphs and travails? They now may fill up all of the backseat of the car or may even be the person behind the wheel, but they still have that marvelous habit of looking straight ahead and sharing their thoughts and asking those sensitive questions. You can have a terrific time planning a tour of a group of college campuses and doing the road trip together.

With the advantage of many years of counseling parents and adolescents on their educational futures, we can share with you a number of useful tips to help you through this stressful experience. We can tell you how to support your son or daughter as he or she confronts the reality of leaving home for college. Try to imagine the emotions churning in your child's stomach as she contemplates the hurdles she faces: getting good grades in junior and senior years as she learns that her performance will define the opportunity to get into a good college; taking hours of timed tests that are seen as a measure of ability; making decisions about interests and educational choices based on little exposure and experience; and, finally, being judged by a jury of strangers who can determine her future!

What can you do to keep the family relationships amicable and your child from falling into his or her own state of panic and despair? Here are some recommendations that have helped the thousands of families whom we have counseled.

LESSONS TO TAKE TO HEART

1. Keep in mind that there is no one perfect institution for your child. No one college can meet the special interests and needs of every young woman or man between the ages of eighteen and twenty-two! Try to get past the rankings as your major source of information and influence. There are dozens of outstanding colleges that can meet a good student's intellectual, academic, social, and personal interests and goals. Focus on what factors matter most for your child, taking into account the criteria we have delineated in Chapter 2 in identifying worthy colleges that may not be household names to you. You will quickly learn that there is a large selection of excellent colleges that will potentially be available to your child. This can remove a good deal of the stress around getting accepted. Any of the colleges in *The Hidden Ivies* will provide the quality of education and growth you would desire for your child, as will many other institutions.

2. Avoid the stampede syndrome. Groups of students, and sometimes their parents, suddenly decide with limited or dubious information what is the hot school of the decade. They treat the one or several colleges as the only place they can possibly be happy. Then they become dismayed that they will have great difficulty gaining admission. Meanwhile they may have missed the opportunity to consider carefully the host of other great colleges that would suit them well. We also refer to this as the bunching effect, which really does result in talented applicants neutralizing one another so nobody wins a coveted place.

3. As the world turns ever more quickly into a knowledge-based economic community and advanced training is required for personal success and security, more specialized knowledge is required. Thus the rise in graduate school enrollments across all fields of knowledge. Undergraduate education, as we describe in Chapter 1, is a step on the road to further advanced education. Those who have researched well and chosen a college that suits their abilities are the most likely to gain entrance into the select graduate schools across the nation. Not only will they gain an excellent and personally rewarding education, they will have the opportunity to explore many fields of study to decide what suits them best.

4. The cost of an education, which defies Newtonian principles of gravity, will continue to rise at an annual rate far beyond the cost of living index and discretionary income for most families. Research the cost of various college options to find the best buys for your child and all potential sources of financial aid and awards to make a college diploma a reality. Cast the application net widely to ensure a good acceptance and aid catch. Keep in mind some of the criteria we have presented: The endowment will determine the scholarship generosity of any particular college, as will its founding mission to offer all or almost all deserving young men and women the opportunity to attend their institution. Accordingly, the top colleges try to maintain a policy of need-blind admissions. Also remember that the so-called tuition sticker price is not necessarily what you will be expected to pay. Academic and talent-based awards are available at the greater number of colleges today. Keep in mind that the majority of financial aid is given on the basis of need. So go for it!

5. Admissions professionals look very positively upon an extra year of high school for the student who wishes to build stronger academic credentials, and in some cases stronger muscles if they are a very competitive athlete, or gain greater maturity. In a number of cases, candidates have spent an interim year working

in a responsible position or volunteering in a community service enterprise at home or abroad. We have counseled many students to act on their desire to travel abroad to perfect their language skills or work intensely with a professional art or music instructor to prepare for advanced studies later on. For anyone considering a year between high school and college, rest assured that you will meet with a favorable response from the decision makers in the colleges.

6. Study the comments of the admissions officers and leaders of each college, like those presented in the individual profiles here, to help determine the rightness of the particular college for your child. A group of colleges may offer similar high standards of education and facilities and resources, but they can attract a certain type of student and deliver a unique style of education that may or may not match with your son or daughter. Note which colleges place a higher or lesser premium on test scores or other criteria in choosing their entering class each year.

7. Reconsider and reconcile your dreams and ambitions for your son or daughter as you work with them through the search process. You are the adult most able to help make a realistic evaluation of your child's abilities and motivation. Overreaching can affect their confidence for the long term, while they can lose out on many good options in the short term. This can lead to frustration and failure if they find themselves in the wrong college.

8. VISIT, VISIT, VISIT! We emphasize every day to families the importance of visiting a group of colleges of differing sizes, styles, and locales in the early stages, preferably taking the initial step onto campuses in the junior year. This is the only way a student can determine for herself what she really feels most comfortable with. High school juniors frequently start with romantic or exaggerated notions of what they want the college years to be like. Some typical articulations we hear: "I am tired of the weather, the people, the sameness of where I have grown up, so I want to go across the country to the other coast," or "I want to go the largest college I can identify because I am in a small high school and I know everybody all too well; my high school athletic teams are terrible so I want to go to a college with nationally ranked athletic teams that I can cheer for; I hate the structured curriculum I have had to take for four years so I want a very progressive college where I can take anything I want and have no requirements to fulfill." Well, you can rest easily if these dialogues arise at home. The great majority of students figure out in time, especially if they have had on-site visits, that there is more to choosing a college than reacting to their high school experience.

MORE ON COLLEGE RETENTION

A number of elaborate studies in recent years indicate that a sense of *social isolation* is a significant factor in students dropping out of college, and sometimes it is the best predictor of failure to complete a college diploma. Other studies have confirmed that emotional and social adjustment to a new environment are more accurate predictors of dropping out or successful completion than academic factors. Confidence in one's abilities is another significant factor that affects success. A self-perception of being over one's head academically or socially can readily lead to failure. Further, interaction both in and out of the classroom, satisfaction with the quality and content of courses undertaken, and a sense of self-confidence affect more positively the determination to finish college.

A commitment to graduating from the particular college and to the institution itself translates to greater persistence to remain and receive the diploma. The obvious factors of adequate facilities, living arrangements, and financial support can affect a student's attitude toward his or her college. A perception of the college as a good place that cares about its students, especially as evidenced by faculty and counselor availability and interest, will have a direct relationship to a college's retention of its students. The inability to cope with adjustment socially and academically results in a greater likelihood of dropping out.

Here is further evidence of the vital importance of choosing the right college. What is considered a good college for one student may be entirely wrong for the next.

PRACTICAL HELP A PARENT CAN PROVIDE

A. Help your child collect all useful resources, such as responsibly written books on the college admissions process, directories of the colleges, college videos and CDs, information from your child's high school regarding recent admissions results for its graduates, and articles from newspapers and magazines.

B. Provide adequate computer capability to allow for browsing the Internet to collect up-to-date information on individual colleges; financial aid resources; special academic, performing arts, athletic, and extracurricular activities of interest; and for downloading and completing admissions applications.

C. Understand how the admissions process works, what is required to qualify, what testing is needed, and what sources of help for counseling and test-taking practice are available in your area. You are then able to gently guide, encourage, and answer questions for your child. You will have to dodge a number of verbal bullets or dark glares along the way, but most of your input will be taken into account in time.

D. Put on your chauffeur's hat, top off the gas tank, and stand ready to get on the road to various college campuses. Most students do appreciate and want their parents to share in this adventure.

E. Keep in mind that one student's opinion does not an entire college make! Be careful of the story of a friend or neighbor's child being unhappy at a college you and your child consider to be right on target. Find out the facts behind the story and judge the college on the basis of all the sources we suggest here.

EMOTIONAL ASSISTANCE A PARENT CAN PROVIDE

A. Share openly and honestly with your child your expectations for college for them. In most cases students believe their parents have unusually high ambitions and goals that they feel will be impossible to meet. This is why many students retreat from the process prematurely with a defeated attitude.

B. Be realistic in helping your child evaluate his strengths, maturity, readiness for college, and the degree of competition and demands he can meet with success.

C. Be the head of the cheerleading section. There are enough critics out there telling your child what she has not done right or why she will not qualify for a competitive college. Be as positive as you can be and leave the yet unmet achievements alone.

D. Learn patience. Your son or daughter will go through a series of fits and starts, optimistic moments followed by crashes as they make their way through the steps toward applying. Their procrastination in meeting deadlines, their mood swings, their failure to communicate with you at various times are all part of the proceedings. They are counting on you to go with the flow, otherwise known as possessing the patience of the saints.

E. Do not make the dinner table the arena for college talk. While this is one of the few places and times when a family sits down together . . . or are we making this up? . . . you can be certain your child will disappear if he finds that the only topic of discussion is colleges and what he has done to move himself forward. Let the food table be a refuge and a comfort from the realities of this stressful time. In its place, set an appointment in another place to go through the schedule of things to be done in an appropriate order.

F. Ask what you can do to ease the pressure and confusion. You might be surprised by the positive response you get when the offer to help is presented in this fashion. It will be perceived as less intrusive. For some students it will be an opportunity to express their self-doubts and worries; others may be concerned with issues of time management and executive skills to get the job done.

G. Share openly your family financial situation. Will financial aid or a part-time job be necessary for your child to attend college? We are no longer surprised by the many students who have no idea if they will need to apply for scholarships, loans, or work study programs. Give them a reasonable idea of how much from savings and income you plan to put to college costs. This will alleviate confusion and make everyone aware of potential financial support from the beginning.

H. Be thee not proud! Yell for assistance from the school guidance staff or, if not available, from professional outside counselors and test evaluators if learning issues are a possible cause of poor test and/or grade performance. A club adviser or athletic coach can often be a good source of information on college opportunities as well as a supporter of your son or daughter when they apply.

I. Take every opportunity to get a good night's sleep. You will need it as you get farther along in the admissions steps. And you will be better able to remember the recommendations you have just read.

J. Live in a state of suspended disbelief through the process! Have faith in your child and his ability to be accepted at a college that truly reflects the hard work he has put into his high school studies and activities. Have confidence that there is a logical, sequential set of steps that will increase his chances of getting into his several top choices. Explore college options beyond the "usual suspects" or familiar names.

FROM VETERAN PARENTS: WHAT WE WISH WE HAD KNOWN WHEN WE BEGAN THE ADMISSIONS PROCESS

1. We thought our son understood that we only wanted him to attend a college where he would be happy and successful, rather than a name college where he might be unhappy while we basked in the glow of its reputation.

2. We thought he understood how much we could afford to pay for his college education; we did not realize it would help to share our family financial situation with him so that he could take advantage of the financial aid opportunities available today.

3. We did not understand that the great majority of private colleges provide larger amounts of financial aid than do public institutions, to make up the great difference in tuition costs. We considered allowing him to apply only to state colleges.

4. We had no idea that college admissions officers welcomed parents to participate in the visits to campus for tours and information sessions.

5. We did not realize that the selective colleges are actively seeking disadvantaged and minority students, that our child could have a potential advantage in gaining admission rather than being discriminated against. All of the recent publicity about affirmative action being done away with is not true for the private colleges.

6. We wish we had known earlier that 80 percent of American colleges and universities offer scholarship grants to talented students regardless of the parents' financial circumstances. We thought only the very needy student could receive aid.

7. We needed someone to explain the differences among all those early application plans and if any of them would help get our child get into a good college. We now understand what early decision and early action and rolling admission plans are and the advantages each can have in admission decisions.

8. We thought that a student could get accepted into a graduate school only if she attended the most prestigious college. We should have known that what really counts is doing very well in good courses in a respected college, that a high grade point average gives her the real advantage.

9. We did not know that major colleges and universities actively encourage students who have demonstrated academic excellence to apply for transfer admission, that a student could start at a college at a comfortable and realistic level for him, do well, and then transfer.

10. We eliminated a number of college choices that our daughter was really interested in because her test scores did not match the average score the guidebooks listed. We now understand that strong grades in top courses and special talents far outweigh the importance of the admissions tests like the SAT.

11. We thought a student had to participate in many different school activities and play a sport. The colleges, we have learned, far prefer the student who has shown major dedication to one or two activities and assumed a leadership role in them. We also saw that only a few terrific athletes were helped to get into college because of their sports involvement.

12. We pushed our child too hard and controlled too much of the admissions process, taking too much of the decision making and responsibility from him.

ADVICE FOR STUDENTS

1. Be conscious of the fact that nothing remains the same in life. What you consider to be your future major interests and goals right now are likely to change as you encounter new ideas and fields of study through your college years. The career you may have in mind today may not even exist by the time you graduate, while new jobs and careers are being created at an exponential rate. In the 1930s, Alfred North Whitehead, one of the great educator/philosophers of the twentieth century, wrote this about change: What differentiated the present from the recent past was a rate of progress that would call upon an individual to face new situations at a pace which found no parallel in history. "The fixed person for the fixed duties who, in older societies, was such a godsend, in the future will be a public danger." What he was saying is that a person trained to do one thing well will quickly become obsolete. His words have more relevance today than ever before. You need to remain flexible and keep an open mind regarding what you study in college and what your future will, in fact, turn out to be. Thus, do not choose a college solely or primarily because it offers a particular academic discipline.

2. Be sure the colleges you are investigating provide a broad range of fields of study and extracurricular activities. Trust your instincts and listen to your inner voice rather than that of your peers, even your friends. You can ask for their thoughts, but ultimately it is up to you to determine what is the most comfortable and best place for you. Be careful about choosing a college because a boy/girlfriend is going there or will be nearby.

3. It is okay to feel anxious as you launch yourself into the admissions game. We would be more concerned if you were not uptight as you join the competition for a place at the selective colleges. The best way to manage the tension is to turn this energy force into action. Get going and do what is necessary, step by step, and watch how much better you will feel.

4. Do not look over your shoulder at what your classmates are planning to do for college or compare their grades and test scores to yours unfavorably. A college of major interest to you can decide to admit you and students with very different profiles if you are all qualified. Decide what you want to strive for and what type of college will let you reach your full potential and simply encourage your friends and peers to do the same for themselves, thank you very much.

5. Start a list of your special talents, accomplishments, and the contributions you feel you can make to a college community. Keep adding to this list as you enter your senior year. You should see that you have more to offer than you probably realized. As you read the college materials pay attention to what characteristics and skills they look for in reviewing applicants and determine which of the colleges match up strongly with your list of strengths and interests.

6. Avoid a focus on names and rumors about individual colleges. Suspend your disbelief, just as your parents should, and instead explore many college choices; make your own preference known, and above all, look for a place where you will find the programs you want to pursue. You can "survive" and attain a good education in many institutions, but discover the one where you will be most comfortable, challenged, stimulated, and successful.

7. Can you afford these colleges? We hope that it has been clear that you should never say never. A clear theme throughout these pages has been the enormous amount of need-based financial aid available at the Hidden Ivies and their great interest in attracting a diverse student body. Apply first. If you are accepted, these schools will try to make it work for you.

HOW TO MAKE IT INTO A HIDDEN IVY: USE THE FIVE P'S

How do admissions officers at the nation's top colleges and universities decide which students to admit from the large pool of outstanding applicants? As you have learned from the comments of the admissions directors and college officials quoted throughout Chapter 3, there are certain factors that count most in identifying students they would like to have in their colleges. One way to understand the admissions decision process and to determine how you can improve your odds of being accepted is to concentrate on what we call the Five P's.

Program

If you are determined to be accepted into a competitive college you should take the most rigorous course load you can handle, starting freshman year of high school. You should tailor a demanding curriculum with a focus on your particular strengths and interests. If you got off to a slower start in the first two years of high school, you can make up for this by taking as many enriched courses in the junior and senior years as you can manage comfortably. You can even use the summers to enhance your academic profile and explore potential future interests. The

Hidden Ivies make it clear that they look for students who have taken advantage of academic opportunities in and beyond high school and who have well-developed intellectual interests and pursuits.

EXAMPLE:

A prospective English major should take Advanced Placement English, literature, and possibly history, but not necessarily Advanced Placement Physics or Chemistry if these are areas of weakness.

Performance

You have now heard from every admissions director at the Hidden Ivies that grades matter greatly as they choose applicants from the mass of candidates. Outstanding performance in a high-level curriculum remains the greatest single factor in a student's chances for admission to an elite college. Serious students should aim for an overall grade point average of 3.5 or higher. Keep in mind that a rigorous curriculum will always show better than a higher average from less challenging courses. It is not necessary to take every honors or advanced placement course your school offers. Concentrate on those that are of great interest to you and that will allow you to excel. Finding a balance is the overriding theme here.

Preparation

By junior year students should know what tests they need to take in order to be considered by their potential colleges of choice, and they should begin making plans to take them at the most advantageous time. Follow this rule: Virtually all selective colleges require the SAT I and three SAT II tests. The alternative testing program that most will accept is the American College Testing (ACT) format. The English with writing and Math level I or II are the two most preferred subject tests. The third subject should be your choice and it should be that which will give you your best result. You may take additional SAT II tests to demonstrate your enriched academic background. You should also check with individual colleges of interest to know if there are more specific test requirements. AP tests are another way for students to show course mastery, and those with a language other than English as a first language will need to take the TOFEL. You should also know if the college programs in which you are interested require particular high school subjects for consideration. If standardized tests like the SAT are not your cup of tea, remember that some of the Hidden Ivies and other colleges have made them optional.

EXAMPLE:

The senior applicant who is short of the required tests is not likely to be considered, as this reflects a lack of planning or lack of qualifications to be taken seriously as a candidate. If in the application you express a serious interest in specific fields of study, you must demonstrate this in your choice of courses in the last two years of high school. An intended engineering or science major should have advanced math, chemistry, and physics; a pre-law major a strong English, history, and economics background; an international relations major a strong language(s) and social studies foundation. And students should have standardized test scores in these areas to help support their applications.

Passion

The elite colleges are in the enviable position of selecting only highly intelligent candidates to fill their classes. But they also want young men and women who pursue non-academic interests with an energy and a passion that make them special. Students who are talented musicians, artists, athletes, community caregivers, school leaders, writers, and so forth make a college campus a rich and exciting environment. This is why students want to attend one of these colleges in the first place.

Parents should be sensitive in the early stages of middle school and high school to those activities in which their child shows special interest or talent and help them to cultivate these. They should help them find outlets whether within their school community or outside it and encourage them to pursue these interests with energy and delight. The smartest course is to drop those activities of minor interest and to pursue those of major interest and enjoyment. The goal is to demonstrate talent, leadership, and commitment. Many students make the mistake of joining a wide range of activities and accomplishing little of note in any of them. Summertime presents a great window of opportunity to strengthen a skill or interest. Here is where research into exciting and relevant programs can be of immeasurable help.

For students, the rule of thumb should always be: Do what you enjoy, and do a lot of it. Yes, you may do some activities because they are required, or are "the right thing to do." But try to avoid involving yourself in something because you think it will look good to a college. And when you find one or two areas that you truly enjoy, pursue them with commitment, dedication, and fun.

Presentation

The personal interview is for many colleges a thing of the past. Few top colleges require a personal meeting with an admissions officer on campus today. They do, however, encourage a campus visit that includes a student-led tour, a group informational session with an admissions representative, and possibly a meeting with a coach or professor in the student's special interest area.

Check with the particular colleges to learn if they do interview on campus, and also consider alumni interviews, which are typically offered once a student has submitted an application.

What has become of indispensable importance to the admissions committee is the personal application. Every top college requires applicants to write several personal statements and include an activity résumé in order to provide insight into their personal qualities and strengths. Most admissions officers place a great deal of emphasis on the personal application. Their feeling is that the essays are very revealing about a student's character, interests, experiences, special circumstances, and, of equal importance, their ability to communicate in writing. We can attest to the fact that many qualified applicants have been rejected by their dream college because of the content or poor quality of their presentation.

You should write about topics that are most meaningful to you, rather than those you think will impress the admissions committee. You should have a trusted teacher or counselor or mature friend read your essays and full application for critical feedback. What is most important is to convey who you are, what the ingredients are that make you the whole person that you know you are and what you will become. Grammar, syntax, and spelling matter greatly, as well, but it is the essence of who you are and what you have accomplished that will make the difference among equally qualified candidates.

It can also be invaluable to send with your application supplemental materials that demonstrate your passions and talents. The admissions committee members are not put off by this. Quite the contrary, they are helped in comparing candidates who profess to have certain talents by evidence of their work and the positive addition this can bring to their campus.

EXAMPLE:

A résumé of theater productions and training, a musical résumé and tape of a performance, a portfolio of art or photography, a bound set of writing pieces, or an athletic résumé and film of competitive play can have a significant impact on the committee. Letters of recommendation from appropriate teachers, advisers, employers, or coaches are also of value.

YOUR CHARGE

The great Irish poet W. B. Yeats described education not as the filling of a pail with information and facts, but as the igniting of a flame. We hope we have made the point throughout this book that the happiest and most fulfilled students are those who experience a faculty passionate about teaching and mentoring, administrators and advisors who care about supporting students in their personal needs, and fellow students who have come to learn in the broadest meaning of this term. These are the key factors that can ignite the love of learning and of exploring new ideas and interests with other talented men and women who inhabit the best of the colleges.

How do you gauge if the choice of college is a correct one? Here is what a parent will want to hear from a son or daughter once he or she is enrolled in college; it is also what a student might expect to be thinking. Parents, don't be fussy regarding the means of communication: Count your blessings if you hear from your child by telephone, e-mail, or even via the back of a napkin from the local fast food restaurant:

1. The work is really hard here, but I like my teachers, especially Professor X in the department I think I am going to major in, because he knows his subject and is enthusiastic and energetic in the classroom.

2. Three out of four of my classes are pretty small so I can ask a lot of questions and also learn what the other kids are thinking. The teachers are cool; the first day they gave us their e-mail addresses and a couple even gave out their home telephone numbers in case we have a question or a problem.

3. I can always get to my professor when I have a problem with an assignment. He has already been really helpful. I know I can always get in touch with all my teachers when I really need to and they will be willing to meet with me.

4. I was invited to assist my science (read any subject here) instructor in the research she is carrying out on a grant. I will have the opportunity to work with her on the experiments she is doing and to help write up the findings.

5. I had a great meeting with my advisor to help me plan my courses for the next couple of terms. This way I can keep a number of options for my major open and meet the requirements for those I am considering.

6. I followed up on my promise and met with the director of financial aid today. He's really friendly and sympathetic to my concern about the number of hours I have to work to pay for all those books I had to buy this term. He is going to look into my aid package and see if there are any extra dollars in their budget to allow me to cut back on my job.

7. I'm impressed by how hardworking and smart the kids are. It makes me nervous, to be honest. But I know I can handle the work and I find it exciting to talk about a book or topic we just covered in class in the student center or the dorm. I'm glad that I took those AP courses last year so I can keep up with the reading and writing they require here.

8. You know what? The admissions representative who came to the high school last year was on target. It's cool on this campus to be smart and study a lot and have a good social life at the same time.

9. I really like the fact that there are so many different kinds of people here. It is so different from high school where most of my classmates seemed the same to me. I am meeting lots of new people with different ideas and opinions, so we can talk a lot about them but no one tries to jam their opinions down my throat.

10. Hey, Mom, the food here sure isn't like yours, but it will do. I have a bunch of options for eating around campus so I get to meet more people and we laugh together about the food. It's the kids here who lift your spirits every day.

11. Well, my roommate is definitely different. Count on that when you visit on Parents' Weekend. But you know, she has a lot of interesting ideas that are nothing like what my friends at home would think. I am learning a lot about her family and their culture. I think we will become good friends over time. We have our differences about things, like what music to play or when to have quiet study time, but we seem to work it out.

12. I'll tell you what is really amazing here. There are so many clubs and groups I can't decide which ones I want to join. I knew varsity sports would take up a ton of time so I am glad I can play on the intramural teams and still be able to write for the paper.

13. You guys will be proud of me. I went to the Career Counseling Office yesterday to find out what I need to do if I want to get into medical school. They had all this great information in a booklet. It tells me what courses I have to take and what exams I'd need to qualify. I also got information on how to pay for grad school and talked with a career and graduate admissions adviser about my options. I think I am going to give it a try now that I know what I have to do.

14. Hey, I hope you won't be upset when you get my final grades for the term. This is a tough place and the professors push you to do your best if you want a top grade. I have learned an incredible amount so far this year in just about every course. I already know what I can do next year to improve. You know, I don't care too much about my grades even though I got mostly A's in high school because I like this place so much and I have made so many friends. Can't wait to see you and thanks for letting me choose to come here.

Now about that car we talked about at spring break. . . .

A CONCLUDING NOTE

We hope that the information we have presented throughout *The Hidden Ivies* has been helpful to students, parents, and others interested in college admissions and in learning about some of the wonderful academic environments offering an outstanding education in the liberal arts today. We would like to encourage students, in particular, to listen to their forebears as they describe their colleges. Look for themes, for the personalities of the schools, and the consistently reoccurring messages. These include calls for your active involvement, participation, going the extra mile, speaking out, taking chances, making connections, and making a difference. The Hidden Ivies are all unique, but they represent top quality liberal arts institutions, each in it's own way spurring students to think critically, to challenge themselves and others, to be leaders. These are communities, families in a way, that invite you in to join them, to contribute, and to take from them knowledge and the means to keep learning throughout your lifetime. These are not places where you can hide, shrink from tough academic work, or values-based decisions. They are not places to float through, where it will be easy to get lost in the crowd or avoid taking responsibility. If you give them a chance and put your heart into any one of these places, you will come out not only a better student, but a better person. Oh, and remember, it's all about the professors, your access to them, and their concern for and ability to inspire you.

APPENDICES]

APPENDIX I

The Hidden Ivies

Amherst College
Barnard College
Bates College
Bowdoin College
Bryn Mawr College
Carleton College
Claremont-McKenna College
Colby College
Colgate University
Colorado College
Davidson College
Emory University
Grinnell College
Hamilton College
Haverford College
Middlebury College
Mount Holyoke College
University of Notre Dame
Oberlin College
Pomona College
Reed College
Rice University
Smith College
Swarthmore College
Vanderbilt University
Vassar College
Wake Forest University
Washington and Lee University
Washington University
Wellesley College

APPENDIX II

Other Colleges and Universities of Excellence

MIDDLE-SIZE PRIVATE UNIVERSITIES

Duke University
Georgetown University
Johns Hopkins University
Northwestern University
Tufts University
Stanford University
University of Chicago
Wesleyan University
Williams College

STATE UNIVERSITIES

University of California, Berkeley
University of California, Davis
University of California, Los Angeles
University of California, San Diego
University of California, Santa Barbara
College of William and Mary
Indiana University, Bloomington
Penn State University
University of Colorado, Boulder
University of Illinois,
    Urbana-Champaign
University of Michigan, Ann Arbor
University of North Carolina,
    Chapel Hill
University of Texas, Austin
University of Virginia
University of Wisconsin, Madison

SCIENTIFIC AND TECHNICAL UNIVERSITIES
California Institute of Technology
Carnegie Mellon University
Georgia Institute of Technology
Harvey Mudd College
Massachusetts Institute of Technology
Purdue University
Rensselaer Polytechnic Institute

Additional Colleges of Note

These institutions meet most or all of the criteria we have set. They have shown marked growth in selectivity and quality of academic and student life over the years. Students report a high level of satisfaction.

Brandeis University
Bucknell University
Connecticut College
Franklin and Marshall College
Kenyon College
Lawrence University
Lehigh University
Macalester College
St. Olaf College
Trinity College (CT)
Trinity University (TX)
Tulane University
University of Richmond
University of Rochester
University of the South

APPENDIX III

Institutional Leaders Questionnaire

We sent this set of questions to the president, provost, chancellor, dean of admissions, director of admissions, and those officials responsible for academic affairs, alumni relations, community services, first-year students, minority students, placement and career services, public relations, residential life, and undergraduate studies, as appropriate, at each institution.

Name _____

Position _____

Years with YOUR COLLEGE?

Graduate of YOUR COLLEGE?   YES _____ NO _____

1.   What are the educational goals of YOUR COLLEGE for your students?

2.   How do you know when you have succeeded in accomplishing these goals with your students?

3.   How has YOUR COLLEGE changed over time?

4.   What has caused YOUR COLLEGE to change?

5.   What helps students to succeed at YOUR COLLEGE?

6.   How do you expect YOUR COLLEGE to change in the future? Where would you like to see the institution go?

7.   Where do your students go after graduating from YOUR COLLEGE? What are they doing?

8.   What programs, policies, or practices have been particularly important to YOUR COLLEGE? Why?

9.   What qualities are you looking for in your students at YOUR COLLEGE, and who does best there?

10.   What are the three most important criteria in the selection of applicants at YOUR COLLEGE?

11.   What kind of cultural, ethnic, racial, socioeconomic, or other diversity exists on your campus, and how does YOUR COLLEGE make itself more accessible and friendly to students with different backgrounds and experiences?

12.   What makes YOUR COLLEGE unique in the world of higher education?

13.   What have we not asked that we should have, and what particular traits would you like families, students, and educators to know about YOUR COLLEGE?

APPENDIX IV

Student Questionnaire

We sent this set of questions to about a dozen students on each campus, sometimes directly through the offices of admissions or public relations.

A cover letter to the students read, in part, "Our goal in surveying students at the colleges and universities we are writing about is to gain an accurate sampling of perspectives from a diverse set of students. We hope to hear from about a dozen students on each campus, representing current opinions from those who know their college well and who can speak to its unique attributes, strengths, and weaknesses. We are seeking sincere and thoughtful responses that will help prospective students and parents understand the "personality" of each environment. Some of those we hope to hear from include: student government leaders, fraternity/sorority and residence hall leaders, international students, multicultural students, athletes, newspaper editors, scholarship students, and so forth. We hope to hear from first-year students through seniors, men and women (as appropriate), and campus leaders."

Name _____
College or university attending_____
Year _____
Membership and/or leadership positions in college or class organizations?
_____
_____

High school graduated from _____
Year _____

Please respond to the following questions as openly and thoroughly as possible, and use any additional space as necessary. Again, you may return the survey to us by e-mail, fax, or regular mail, as works best for you. We will use the information we gather to discuss your school in depth, and with the benefit of your knowledge and insights, we hope to capture its unique qualities. We will report responses in our book anonymously. We appreciate your time and your candor, and we value your contribution to this discussion of excellent institutions of higher education.

1   Why did you choose your college?

2. What does the school do best?

3. What does it do worst?

4. Who does best at your school?

5. What helps students to succeed there?

6. What would you like to change about the college?

7. What are your educational goals?

8. What programs, policies, or practices have been particularly important to you at the college? Why?

9. What do you plan to do after graduation?

10. What kind of cultural, ethnic, racial, socioeconomic, or other diversity exists on your campus, and how does the college make itself more accessible and friendly to students with different backgrounds and experiences?

11. What makes your college special?

12. What haven't we asked that we should have, and what would you like families, students, and educators really to know about the school?

GREENES' GUIDES
TO EDUCATIONAL PLANNING

Making It Into a Top College

ISBN 0-06-095363-2

This definitive step-by-step plan for organizing the admission campaign reveals an inside look at how the admissions process works at the most competitive colleges, what the top colleges are looking for, how to choose the best college for you and much more.

Making It Into a Top Graduate School

ISBN 0-06-093458-1 (Available August, 2001)

A ten-step strategy for finding and getting into the right graduate schools, complete with time-tested worksheets, executive task checklists, sample essays and resumes, inspirational case histories, interviewing techniques, and program categories and rankings.

Presenting Yourself Successfully to Colleges

ISBN 0-06-093460-3 (Available August, 2001)

A step-by-step guide to help students market their strengths and uniqueness in an outstanding college applications, including essays that will be noticed, targeted interest resumes, portfolios, and letters to coaches and instructors. Complete with actual samples from the files of thousands of the Greenes' clients.

The Hidden Ivies

ISBN 0-06-095362-4

An in-depth survey of thirty colleges and universities of exceptional merit that will not only challenge gifted students, but also provide them with the competitive edge needed upon graduation. This book distinguishes each college's personality as described by students and campus educational leaders, as well as its unique academic strengths.

The Public Ivies

ISBN 0-06-093459-X (Available August, 2001)

An excellent resource providing a better understanding of the differences between various public universities, the contrasts between public and private higher education, the admissions process for public schools, and what it takes to succeed in the larger, public university environment.

Inside the Top Colleges

ISBN 0-06-092994-4

The quality of education and quality of life at the country's 20 most elite universities are explored, based on a survey of 4,000 undergraduates using advanced polling techniques, extensive focus groups, and one-on-one in-depth interviews with students.

Available wherever books are sold, or call 1-800-331-3761to order.